A Beginner's Guide to Version

System Excellence

Vinay Singh

Ph.D.(Computer Sc.), M.S. (Computers), MTech(I.T.),B.Sc.(Computers)

Colorado, USA

Rakshit Singh

Bachelor of Science (Software Engineering)

Toronto, Canada

RoseDog Books

PITTSBURGH, PENNSYLVANIA 15238

First printing edition 2023.
Vinay Singh (Author)
Email: usavinaysingh@yahoo.com
LinkedIn: https://www.linkedin.com/in/vinay-singh/
Google Scholar link
https://scholar.google.com/citations?hl=en&user=PT_e528AAAAJ
Ph. No: +1-973-906-3275(USA)

RoseDog Books
585 Alpha Drive, Suite 103
Pittsburgh, PA 15238
Visit our website at *www.rosedogbookstore.com*

ISBN: 979-8-88925-399-0
eISBN: 979-8-88925-899-5

Table Of Contents

Acknowledgments

Writing a book is harder than I thought and more rewarding than I could have ever imagined. None of this would have been possible without my family. My Mother has been source of inspiration for me.

I'm eternally grateful to my friend Shailendra. He taught me discipline, tough love, manners, respect, and so much that has helped me succeed in my IT career. I truly have no idea where I'd be if he hadn't given me a roof over my head or become the father figure whom I desperately needed at that age.

To Prof. & Dr. Alok Aggarwal, who inspired me to start my research work. He helped and encouraged me.

Writing a book about the story of your life is a surreal process. I'm forever indebted to lovely Vedanshi, Rajni and Rakshit for their love and happiness in bringing several stories to my life. It is because of their love that i could write the book in a professional style. A very special thanks to them.

To everyone at the Michelle who enables me to be the Sr Manager DevOps practice of a company that I'm honored to be a part of, thank you for letting me serve, for being a part of our amazing company, and for showing up every day and helping more authors turn their ideas into stories.

Finally, To my family. To my Father-in-law Mr. Amar Singh for always being the person I could turn to during those dark and desperate years. She sustained me in ways that I never knew I needed.

.

What readers are saying about

A Beginner's Guide to Version Control System Excellence

Natasha Taylor

IT Specialist (Germany)

If you are thinking of learning Git, I highly recommend this book. If you are not using a version control system. You'll find distributed version control and Git much interesting after reading this book.

Joseph Williams

Sr. Manager Software Development (USA)

Not only has this book convinced me that Git has something to offer over CVS and Subversion, but it has also showed me how I can benefit from using it myself even if I'm using it alone in a CVS/Subversion environment.

Adams

Senior DevSecOps Engineer(USA)

Mr. Singh has done an excellent job taking a tricky subject and making it accessible, useful, and relevant.

Stanley Peoria

Entrepreneur(Canada)

This book is amazing. Thank you Vinay for teaching me version control and GIT concepts. Excellent Job.

Soumya Verma

Student ,(Kenya)

Thanks sir, I got highest grades in my subject.

Preface

Agile and Kanban in modern development methodologies has changed the way software development world and business deliveries. The whole efforts are given to migrate from old style traditional centralized version control systems (VCSs) to a lightweight, distributed version control system (DVCS).

As business continuity is an essential requirement in every aspect especially in Real Time mission critical applications like in financial, defense, banking and space domain etc.

Git has been around since 2014 and developed as advanced and modern Version Control system(DVCS). Its following strengths can help you as a git administrator or expert in DevOps and Software development.

Who's This Book For?

This book has been designed for everyone from every domain. If you are a Student in university or a DevOps engineer or a software programmer, this book will help you in all dimensions. If you're new to version

control, this book will walk you through thc basics and gct you up to speed on how Git can help you.

If you're a software developer and worked with Subversion or other traditional tools, you can still learn several advanced topics.

What's in This Book?

This book is divided into 13 chapters.

Chapter 1: Introduction to Version Control System

Chapter 2: Version Control System Terminologies

Chapter 3: Centralized Version Control System

Chapter 4: Distributed Version Control System

Chapter 5: Installation and Setting up GIT

Chapter 6: GIT Commands

Chapter 7: Branching and Merging Strategy

Chapter 8: Security in GIT

Chapter 9: GIT TOOLS

Chapter 10: Advanced Topics

Chapter 11: GIT Administration

Chapter 12: SVN to GIT Migration

Chapter 13: DevOps and GIT

Chapter 1
Introduction to Version Control System

A version control system (VCS) is a special kind of software or tool used to track a file's changes and gives you a report on who made the changes when the changes were made and what were those changes.

1.1 What is Version Control System

The difference in the contents of a file at two different times is called a version.

Consider yourself in a complex software development environment where you are leading a team of several developers and suppose to initiate a new software project. These developers will be working on several modules of that project. This project needs to be segregated into smaller pieces i.e., some logical modules to meet business demand. You need to meet the project deadline. So, you assigned the objectives to each developer and its respective module. Developers are given project end dates and they rigorously work on writing complex code to meet module deadlines.

Let's take a scenario in which each developer is given an independent module and is expected to meet deadlines. They make changes to their code on a day-to-day basis or sometimes several times. They add some code or sometimes delete the portion of code to fix the bugs coming in their module. After some time, they will need to refer back to the previous code written by them or you need to track their progress or review the code.

Let's talk about the second scenario where some modules are dependent on each other i.e., in order to execute them the parent module needs to be executed first, and sometimes they have a common code or common functions which are indirectly calling other functions. In this case, sometimes developers are required to make some modifications to other developers' code so that all modules can run properly.

This is where a Version Control System (VCS) comes into the picture to help in tracking the several versions of the changes made into single or multiple files by single or multiple developers.

Need of Version Control System

The Version Control System is a software tool that tracks the changes made to a file or a set of files and takes a snapshot of every single minor modification.

This software helps the Software development teams save their coding efforts in an organized fashion so they can fetch their code and build the customer solution whenever needed. Each version of a file committed at any point in time is recorded for future audit or review purposes.

The definition and requirements for the version control tools are very clear to us but still, an important question arises why Software Industry should heavily invest in these kinds of tools? What are the direct and inherent advantages of a Version Control System that businesses can avail?

Version control is beneficial by keeping a record of all kinds of version code like text, binaries, and blobs kind of digital assets. It plays a crucial role in software development culture by managing and keeping track of developer code. Collaboration and communication are two major principles on which any Version Control System is built and designed.

When we leverage the Version Control System features into Software Development Life Cycle(SDLC), it helps us to organized the source code in better manner. There are several features generally a VCS possess and helps development team to keep their code and its versions at one single place.

The following are some high-level features of VCS:

- Single Source of Truth
- Traceability
- Branching and Merging
- Reduction of Duplication and Errors
- Compliance
- Time to Market: CI/CD Automation
- Policies and Hooks
- Collaboration and Communication
- Security

Single Source of Truth:

A Version Control System becomes the single source of truth because it accepts all the changes and puts these changes in form of several files at one accessible location, This single location with those set of files that represent the developer team's assets is the single source of truth. The whole development effort revolves around the Version Control System.

In parallel processing, If developers write the code concurrently, version control keeps them in one place and allows them to quickly revert changes to a previous version in case the software build fails or there is a severe bug is found. Version Control System also facilitates the comparison between two versions of files or identifies the user or commit time through the revision history.

A developer is always expected to push their changes to Remote Version Control System to be hosted remotely so that the whole development team can access the full project from one single place.

In essence, a VCS is a truly strong candidate for the source of truth to keep and maintain the files.

With a strong VCS, all the stakeholders like development, Testers, Release Engineering, Product Owner, and DevOps teams can effectively communicate and better collaborate to stable build innovative solutions. Hence it is essentially a Single Source of Truth.

Traceability:

A Version Control System supports tracing back to the changes made by several developers on several files. VCS has control over all the software projects and repositories to track back the history of the changes made in the project. It also facilitates the developers to fetch the metadata of the data and the datasets in the projects. With the help of VCS now developers can seamlessly commit several changes in the remote repository. The traceability feature is essential to any software development life cycle[SDLC].

VCSs are also called SCM or bug tracking systems to annotate each change with a message describing the purpose and intent of the change.

For example, git status or git log helps to track the changes in GIT and the same for SVN as svn log or svn diff can be used to trace the code and find history.

There is a special type of object is used in GIT, The tree object is a way that helps Git to keep track of file names and directories. It means that a tree object has to be associated with each directory. At the time of the commit, It is very imperative and essential with legacy code to estimate future work and find the accuracy in the code itself.
The SHA-1 blobs, the files, in that directory, and the SHA-1 hash, sub-directories are core dimensions of tree objects.

Being able to trace each change made to the software team and project management, VCS is very helpful to support traceability requirements in SDLC culture.

Branching and Merging:

A branch is fundamental means of creating a separate line, that is available and managed into a version control system (VCS).

It facilitates the developers to continue their development work within a software project. This eventually helps to software development teams work simultaneously.

A VCS inherently establishes a relationship between the newly created branch and the parent branch it originated from.
By working on separate branches, developers can produce different versions of the software project.
A VCS then allows several parallel branches to coexist and then merged on a demand basis with other branches to have all common features and efforts by several developers. It means Branching entails the DevOps principle to collaborate and communicate inside of one central code base. A copy of the code base is created instantly as a software developer creates a branch,

DVCS like Git allows several branches and thus various lines of development within a repository.

Merging is the process of combining the efforts of several developers working on parallel branches and then integrating their changes.
It means that Merging is combining two or more different branches and their files including the functions, variables, and their history as a single unit into one.

Code branching and merging are how developers work on changes and merge them back into the mainstream.

A VCS like GIT hub, Git Bitbucket, or any other VCS compares the two distinct sets of changes and track to determine where those changes occurred.

Reduction of Duplication and Errors

Redundancy means duplicate or similarities in a code that has been produced by group of developers by producing multiple versions or artifacts of a file or portion of the code. Code reuse is the best practice in SDLC but there must be careful analysis and attention need to be given if developers are rewriting the same code or reusing the same code in their functions or libraries.

Version control facilitates the reduce the duplication of multiple and outdated versions of the code written by developers. I

First of all, the bad impact of redundancy is unnecessary increase in software code volume and number of lines. Every supplementary line of code will, sooner or later, enter into the maintenance process, and thus cost time and money.

Secondly, duplication directly influences the difficulty of maintaining already-developed code. A classic example could be that developers trying to modify a piece of duplicated code.

Reducing the duplication in code directly entails reducing errors that has been arise as sub-products in various multiple documents. sometimes it causes Merge conflict issues.

Many advanced version control systems give you some way to identify the files(blobs) and their copies by the indirect use of commands or they allow you to integrate with third-party duplication detection tools. This duplication detection tool must have capabilities to detect duplicates in presence of such changes keeping a balance on what to be removed or what not to be.

This generates a recursive list of all 'blobs' in the current HEAD, including their path and their sha1 id. We will discuss it later.

Here are some points of consideration to avoid the situation of duplicated files, we should never reuse code by copying and pasting existing code or a portion of the code. Rather, a careful strategy should be applied to put it in a method if it is not already in one and call it when needed.

Compliance

Every IT organization strictly focuses on having proper transparency and clear visibility into their software changes. Tracking who, what, when, and why changes are made is important.

There are several use cases for example some regulatory bodies like HIPPA or FDA, governance, risk, and compliance (GRC) where a system needs to stay strictly compliant. Version control software captures and controls the detailed information of who made changes, who can view the changes, when the changes were made and what were those changes. It also maintains the software project history forever for future auditing and compliance reports.

Document version control allows its all-team members to save, view, and organize important compliance information and lets them easily access relevant documentation using authentication and authorization.

FDA, HIPAA compliance software that helps compliance officers with security risk assessment and incident management.

VCS provides the ability to control who can access documents and the ability to keep a detailed record of who makes changes and stay compliant with regulatory bodies.

Time to Market: CI/CD Automation

Time to market is biggest demand of today's business world. It is considered as the total time taken by the development team in releasing a software product to the production server as a final business delivery to the end customer and market. This can be due to analyzing, designing, coding, building, testing the code at various level, and Since research has shown that a business always seeks to get new products to excel in its dimensions in terms of market share, revenue, and sales growth, time to market is one of the essential products KPIs or metrics. Therefore, businesses always strive to enhance their product features to market more quickly, so they avail all possible opportunities and easily adjust to the market standards.

When development teams build software, they reside their code into a Version Control System and then build this code for testing. If the product is to be developed by developer first time then, operations need to setup a fresh ready quality assurance and testing environment for several types of testing like integration testing, functional testing, regression testing and performance testing. Hence, a developer can only deploy code to the software when it is available in some repositories and ready for build and deploy a team to build and send communication to the software testing team to test the product.

Every organization seeks its product owners to send new features faster and faster. Here developers have to be proactive, flexible, and agnostic at the same time to adopt any advanced VCS of their need or any programming language with the capability to build the final product. Here version control plays a key role in DevOps that helps in reducing time to market with the integration of a modern DevOps tools like Jenkins, Kubernetes and Junit etc.

Collaboration and Communication are essential part of a development efforts and entails to be operated more proficiently. To achieve these objectives, organizations must adopt the right DevOps practices in their software development processes to reduce time to market.

Hence, Integrating VCS and DevOps CI/CD practices can make faster time to market.

Policies and Hooks

Build automation is the process of automating checking out source code as soon as the developer makes some commit in the Version Control System, then calling a CI/CD pipeline to compile, running some automated tests to test unit test cases, and publishing the version of this new build at some location like Nexus and publishing it into a shared, centralized repository finally deploy this to the respective environment.
The first step in Build automation is committing the final code to VCS but we need to make sure that developers are following the agile process's local policies.
This can be achieved using "Hooks".

Hooks are implemented to address the local policies and execute some actions on some events. It means they are purely event-based. The hooks reside in a directory within the VCS repository to see if there is an associated script to run.

Hooks are generally implemented in scripts that improve functionality and enhance the restriction policies based on different events and actions.

Some examples of hooks are here

pre-commit: It occurs before a git commit is accepted.
post-commit: It occurs immediately after a git commit is accepted.
post-checkout: It occurs before a git checkout.
pre-rebase: It occurs before a git-rebase.

pre-commit hook:

It occurs before a git commit is accepted. For example, a restriction or policy can be applied for developers to make sure they mandatory mention their Jira story number in a commit message.

While the advantages of git hooks are comprehensible, it is cumbersome and complex in their actual implementation.

post-commit hook:

This hook is invoked by git-commit. It does not take any parameters. It is invoked after a commit is made and usually calls a CI/CD pipeline to support build automation.

Collaboration and Communication

When you work with a version control system anyone on your team participates and collaborates with the role provided. VCS administrator gives permissions to several team members based on their role. Some users can only read the files, some of them can edit the files, and some of them can review and accept the changes or merge requests.

A VCS facilitates collaboration and communication among several team members geographically separated and provides a single platform, which ultimately increases the team's efficiency and output.

An agile team can appoint a scrum master or lead developer to do a code review of the code written by several developers. Developers send pull requests i.e., to merge their feature branch then a code reviewer can review the code and check and revert the duplicate, unethical, malicious code.

Now developers can more easily create branches and merge their changes to the main branch. In parallel processing culture, developers work on their feature branches independently without blocking other team members. They send the pull request to the code reviewer to merge their feature branch to the development branch or merge the bugfix branch back to the release and development branch. They communicate it by sending a commit message in the pull request. A code reviewer or group of code reviewers by extensive means of collaborations and either approve or reject the pull request.

Some VCSs also facilitate collaboration via tools to share edits with others and resolve conflicting edits.

A code reviewer or collaborator perform these responsibilities:

- Approve the pull requests.
- Publish and install the packages.
- Deny the pull request.
- Allows to make change in the code as suggested by the Pull requests.
- Mark issues or pull requests as duplicates.
- Adding others or removing themselves as collaborators.
- Approve the releases in the repositories.
- Fork the repositories.
- Edit commits or pull requests.

Security:

A vulnerability is a severe flaw in the system or software that can be exploited by hackers snooping around on the internet to gain unauthorized access to a system.

Vulnerabilities are increasing day by day in every aspect of IT resources and Software. The efforts given by developers to make a product better are wasted when their code is found with vulnerabilities. It has been seen that several VCSs are no longer safe and are said to be vulnerable to attacks and flaws. Cybercriminals tricks and execute a code on the victim's platform and steal sensitive information.

But advanced Version Control systems like GIT, allow you to use SSH keys as a means of proving your identity.

Another advantage of version control is that they are distributed and sometimes developers keep their copy at the staging area. In the case a cyber-attack or hacking happens to their system, the other developers have a copy as well.

It is a best practice that developers should carefully keep sensitive files with VCS's ignore functionality.
In an advanced version control system, The git-secret utility can store sensitive data inside a git repository.

$git-secret-init - initializes a git-secret repo by setting up a .gitsecret directory.

This is similar to the git init command. It creates a new .gitsecret directory inside the repository.

Key Vault increase the security and apply several polices to keep the code as secrets which might be accidentally leaked. With the help of a key vault, developers are not required to store security information in their applications. A key vault can be easily integrated with advanced VCS.

The reasons for security leaks or vulnerabilities could happen due to hardcoding sensitive information, misconfigured access permissions, adding passwords in files listed in the .gitignore file, Unprotected branches, Git vulnerabilities use security by GPG key.

Some advanced level VCS Security Best Practices are suggested:

- Avoid storing credentials in VCS.
- Create a SECURITY.md file.
- Disable forking by VCS admin.
- Disable public repository creation and commit to them.
- Use GitHub teams for access control.
- Configure IP whitelisting.
- Apply MFA tokens.

1.2 History of Version Control System

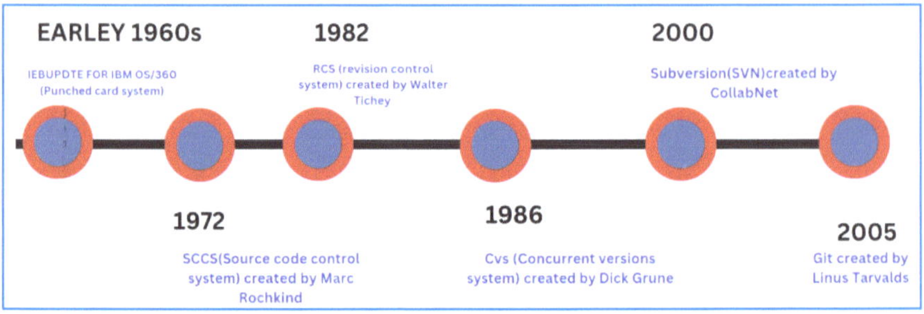

Figure: History of Version Control System

Before we start using today's sophisticated Version Control Systems like GIT Hub, Git Bitbucket, or TFS, it is suggested that we should know the background of the evolution of VCS.

The original version control software was mainframe-based, and these VCSs had limited features. They used to provide some interface via a terminal. Then centralized version control systems were introduced with a single, shared repository.

The first generation (1972-1982)

The first Version Control System was developed in famous Bell Labs where they developed UNIX in 1972. It was SCCS(Source Code Control System), a file-centric and local system. It only worked with Source Code files.
The basic operations like deleting, renaming, or creating a file were tedious and required lots of time and effort. There were several problems in that system as tracking files,
branching and merging and Locking a file, and parallel processing. features were missing.

In 1982, Walter Tichy introduced a new system RCS(Revision Control System). It also had some limitations like it allowed one user at a time to make edits. It means that another user could not work on another file on the same project.

The second generation (1982-2003)

In 1986, Dick Grune presented a paper with a new concept of CVS (Concurrent Versions Systems) and developed this software. CVS finally allowed more than one developer to work on a file at the same time. By that time C language was very popular hence Dick used C to develop it.

The next major version control system was Subversion (SVN), developed in 2000 by CollabNet. By 2010, SVN was renamed Apache Subversion after it became part of the Apache Software Foundation. Some of the most popular VCSs in the second generation were SVN, CVS, ClearCase, and Perforce.

The third generation (2003-present) :

The third generation was distributed version control (2003-present) – It provided first time in history a merge-based distributed and concurrency model.

The era came for distributed models. The DVCS(Distributed Version Control System) made the process of forking, branching, and merging very simple and reliable. The basic pillars of DVCS were Remote repository and t a copy of the repository and its history is available to every user at their local. Currently, the most well-known DVCS is Git, which was created in 2005 by Linus Torvalds. The model of Distributed Version Control Systems (or DVCS) became mainstream with BitKeeper, followed by Git and Mercurial.

Let's have a glance at some popular VCS

Concurrent Versioning System (CVS) – 1986. CVS introduced the concept of a repository. While RCS had a concept of file locking, CVS loosened the restriction and defaulted to a merge model to resolve conflicts. It also was one of the first VCS to introduce branching and symbolic tags.

ClearCase – 1992. Atria company introduced ClearCase. One of the first commercial VCS. ClearCase emphasized networking capability in VCS as it allowed to do storage operations from the local disk to networked file storage, which later merged with Pure Software (founded by Reed Hastings, who later founded Netflix).

Perforce – 1995. One of the first VCSs introduced the first time branching and merging concept.

Subversion – 2000. Subversion was the open source that introduced client/server architecture. VCS. Some inherent features of this VCS were but not limited to atomic commits, branching, merging (and file locking), binary file versioning, and client/server architecture.

TFS- Microsoft developed this long back and now it is called Azure DevOps. Team Foundation Server was first released on March 17, 2006.

GIT- 2005. The third generation of VCS was distributed. The next source code control software generation is called Distributed Source Code Control. Distributed Source Code Control is different in that there's no central server. It means everyone keeps all files locally and commits the changes to the remote server whenever needed. It allows offline building which makes it a strong candidate for modern VCS.

1.3 Types of Version Control Systems

Version control systems are specialized software tools that help to track modifications done in the code. Version control not only maintains the version of software code but provides a complete history of every file. As we know that they have evolved over the years. They have been improved and architecturally modified over the decades.

There are 3 types of Version Control Systems:

- Local Version Control System
- Centralized Version Control System
- Distributed Version Control System

Local Version Control System

It is one of the simplest forms. A local VCS uses a revision control methodology to keep all the changes to files using a large database. RCS is one of the most common VCS tools. The Local Version Control System calls this database a version database. It stored patch sets. A patch set is a difference between files. By the use of these patch sets RCS recreates full past versions of the file. Here patch sets are kept in a special format on disk.

Example: RCS

1.4 Benefits of using Local Version Control System

- Simple
- Easy to understand
- Tracks change over the period

1.5 Disadvantages of using Local VCS

- Single Point of failure
- Weak collaboration

Local Version Control System is prone to error and very unsafe as it stores the files locally. generally speaking, that Local Version control system are not considered for team projects in modernized culture.

1.6 Centralized Version Control System

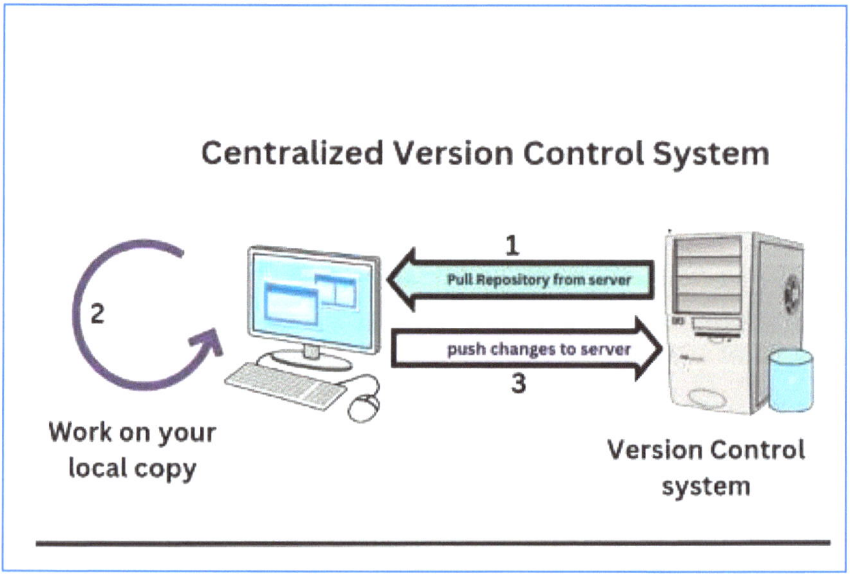

Figure: A glance at Centralized Version Control System

The architecture of Centralized VCS is different from Local VCS as it keeps only a single "central" copy of your project on a remote server and allows developers to commit their changes to this central copy. u need, Some restrictions are applied as you cannot have a full copy of your project, but you can pull the files.

It's very similar to FTP in that you have an FTP client which connects to an FTP server.

Collaboration was the central principle behind the design of Centralized architecture to allow several teams to collaborate easily. It means multiple users can work with the same files concurrently. Several developers from different teams can check out and edit different files from a remote server. This system became the standard VCS setup for many years.

Examples: CVS, Subversion (SVN), and Perforce.

1.7 Benefits of a Centralized Version Control System

- Collaboration
- Comprehensive
- More Administrative control
- Simple to install and configure
- Tracking is easier than Local VCS
- Enables Collaboration
- Administrative Control

Centralized Version Control System has several benefits like it is reasonably easy to set up, it permits us for file sharing amongst developers, teams can pull a few lines of code without saving the entire history on their local machine and it provide us various options, all projects are stored on a more reliable server like datacenters or cloud. On the top of everything , VCS Admin can control the access and structure of the repository.

1.8 Disadvantages of a Centralized Version Control System

- A single failure can drastically software code loss forever
- Hard to manage a server and backups
- Branching and merging are difficult.
- It can be slower due to the network issue

1.9 Distributed Version Control System

Distributed version control systems (DVCS) is completely different from Centralized VCS as keep all the project locally and have full control of all the versions of a project's files. DVCS allows you to clone and read the full history of the project.

It means If any server crashes, any of the client repositories can be copied onto the remote server which supports continuous development efforts.

There is a special area in DVCS architecture which is called the "Staging Area"

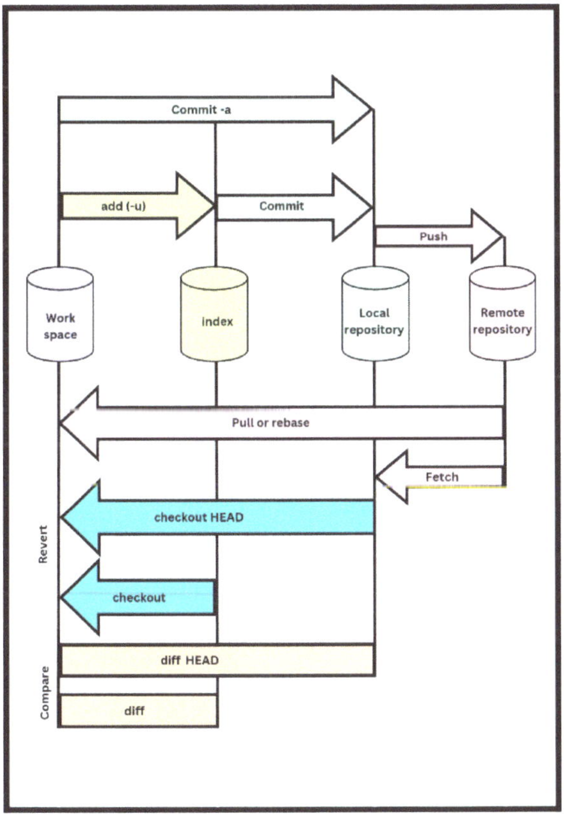

Figure : A holistic view of DVCS workflow

commits, proactively reviewing changes and learning in advance if the merge has code conflicts.

Indexing is a very special technique in GIT. It facilitates the developers to commit only certain parts of the changes they feel to move forward with. Say you're working on two features - one is finished, and one still needs some work done.
Let's take a scenario where a developer is ready to commit for a feature but does not feel comfortable to commit the other parts of the second feature and which is not finished yet. Now developer stages the parts belong to the first feature, and commit. It means the files which are staged and indexed are only committed and pushed to the remote server. Isn't a fantastic feature?

Example: Git and Mercurial.

1.10 Benefits of Distributed Version Control System

- Merge conflict resolutions is easier
- Less dependency on the remote server
- Branching and merging are easier and more flexible
- Better collaboration
- Support offline builds.
- provide a staging directory for indexing

1.11 Disadvantages of Distributed Version Control System:

- Complex architecture
- No incremental revisions

Chapter 2
Version Control System Terminologies

It is very likely to understand the terminologies of the Version Control System.
Below are a few very common version control terminologies before we deep dive into the VCS ocean.

2.1 Versioning

Versioning is a technique which tells the unique states of code at any point in time which is developed and released in organized fashion by a VCS. A version identifier is a word, a number, or both. For example, version 1.0 can be consider as the first draft of artifact ready to be released of a program and 2.0 could be the second version.

2.2 Repository

A repository is a fundamental unit of VCS that stores all the files and changes made by the developers to their projects. It is useful for a team of developers it can be especially useful as they can keep their project and its associated files separated from the other developers from a different team. It is a .git folder inside a project, There are several repositories in a VCS.

2.3 Project

A project allows you to group repositories and manage permissions for them in an aggregated way. Each project contains at least a repository. Projects can be available publicly, internally, or privately. A Git project consists of 3 sections: Git directory, Working tree, and Staging area.

2.4 Branch

Developers create branches to get alternate versions of their projects. When a developer gets a new assignment that is different from existing product functionality then he gets new ideas to implement this. Finally, developers create branches so that they can work on new features and try to compile the code without messing with the main project work. Usually, there are four kinds of branches, Master Branch, Develop Branch, Release Branch, and Feature Branch. Branches exist inside a repository.

2.5 Commit

When developers commit a change, they add their latest changes to their branch residing in its repository. When you commit a code from local, you actually not only push your changes to the repository but its meta data and history as well.

2.6 Head

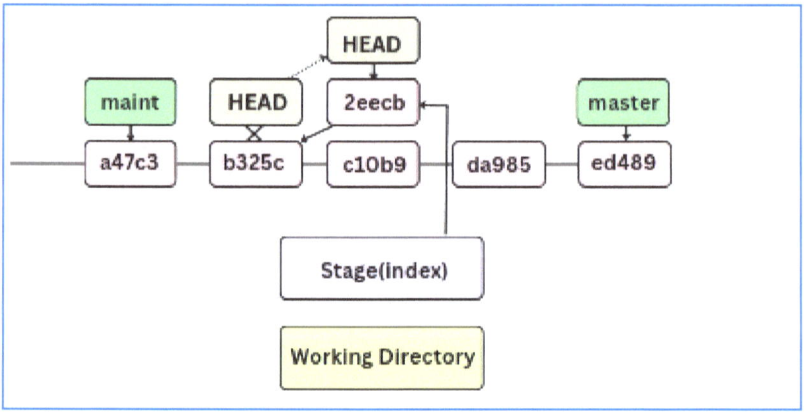

Figure: A Head in a branch

A Head is a pointer to the most recent commit in the current branch. This means HEAD can be consider as a reference that keeps track the current branch and its latest commit in that branch.

2.7 Diffs

Diff is used to identify the difference between the two commits by comparing the two versions. Every time a developer add something to its code and commit, there are differences added into metadate about the last saved point to the new one. These changes are called differences, or "diffs" is used a short term.

2.8 Merging

Merging in VCS is combining two separate development histories into one. When the developer executes the git merge command, it allows them to join one branch code changes to another git branch and integrate them into a single branch. In a very simple way, you can merge code from a separate branch into the main branch to integrate the changes.

2.9 Conflict

A conflict is a situation which occurs when two developers make change on the same file or same line of code. Merge conflicts happen when two developers work on single a single file without syncing from the remote repository. Pulling the code from a remote repository helps to avoid merge conflicts.

2.10 Clone

Cloning is a way to create a local copy of your repository to start development efforts locally.

2.11 Pull

A developer must need to sync with all other changes pushed by several, other developers. Pull request allows syncing all the new revisions of the remote repository with the local repository.

2.12 Forking

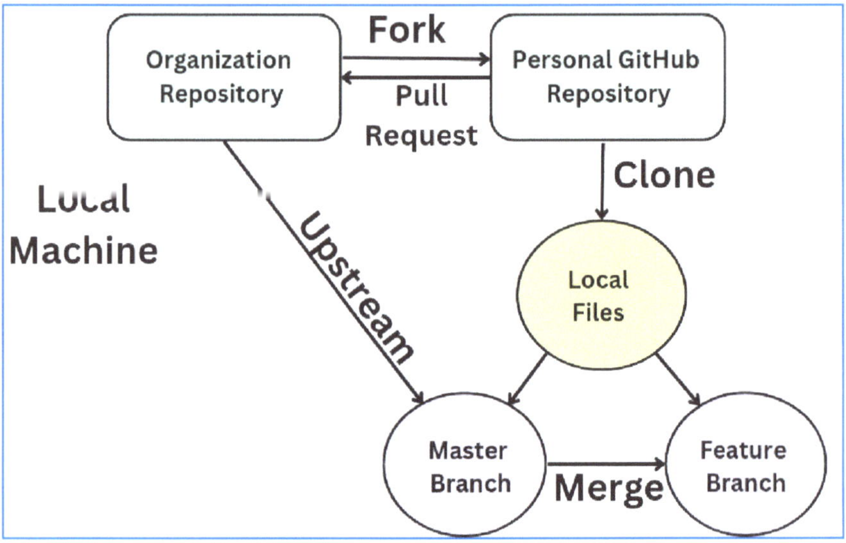

Figure : A fork command and its impact

Forking is just like a fork in a tree. A forking involves making a new repository, It means, a fork keeps everything of the original branch, but a developer who forks will be the owner of this new repository.

2.13 Checkout

Check out, lets developers switch from one branch to another branch at local working copy. Sometimes it is used to create a new branch if a branch is not created and then switch to this branch.

2.14 Working Directory

The Working Directory is the location where you clone your project from a remote repository.

2.15 Staging Area / Staging Index / Index

Index will maintain Tree and blob of the objects. Index is something that acts as interface or area between local repo and working directory. At this stage developer can easily decides what needs to be sent to the local repo. it decides what needs to be sent to the remote repo and what needs to be kept as it is.

2.16 SHA

An SHA is "SHA" is shorthand for "Secure Hash Algorithm". It is a unique ID number for each commit made by the developer.

A typical SHA might look like
"*k2rcn889e7e26ged4skby75tn44aad11908kj55v*"

It is a 40-character string (0–9 and a–f) which is combination of characters. It is calculated based on the contents of a file or directory structure in Git.

2.17 Tag

Tags help to mark an important commit and identify which can be recognized or used in later stages like releasing a project with a specific tag number. These are special reference numbers showing particular points in a Git history. The basic utilization of using a tag is to capture a point in a Git history that marks a version release. Keep in mind that Tags don't change. After a tag is created, it has no history of commits.

2.18 Working Tree

A Git work tree is a special kind of data structure, allowing to have multiple branches checked out at a time. Actually, it is a linked copy of your Git repository.

2.19 Directory

A directory is where VCS usually stores the metadata and object database for a software project.

2.20 File

A file is generally referred to as a piece of information that has some specific meaning and relevance to the project. For example, source code files like java or python files, and configuration files like README and LICENSE files. Some placeholder files .gittkeep is used to add a placeholder for an empty repository.

2.21 Versioning Naming Convention

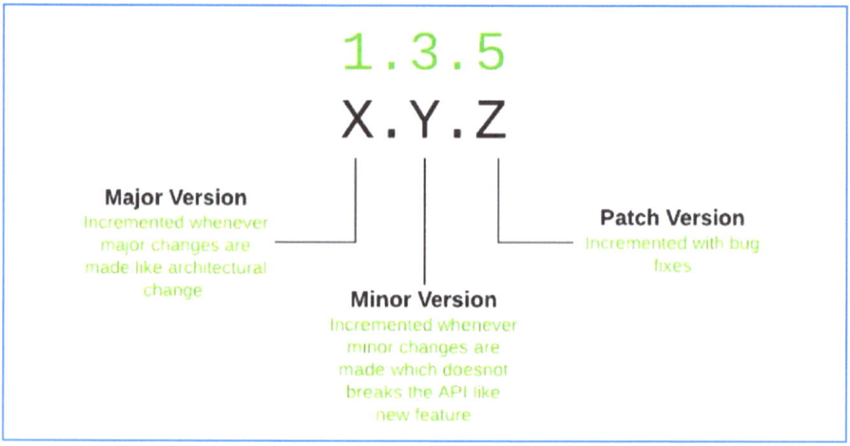

Figure: Conventions in versioning

Software industry usually use Semantic Versioning. it is a 3-component number in the format of X.Y.Z, where :

- X stands for a major version. major change in product.
- Y stands for a minor version. It denotes the release of new functionality.
- Z stands for a Patch Versions: It denotes a patch is used for bug fixes.

Chapter 3
Centralized Version Control System

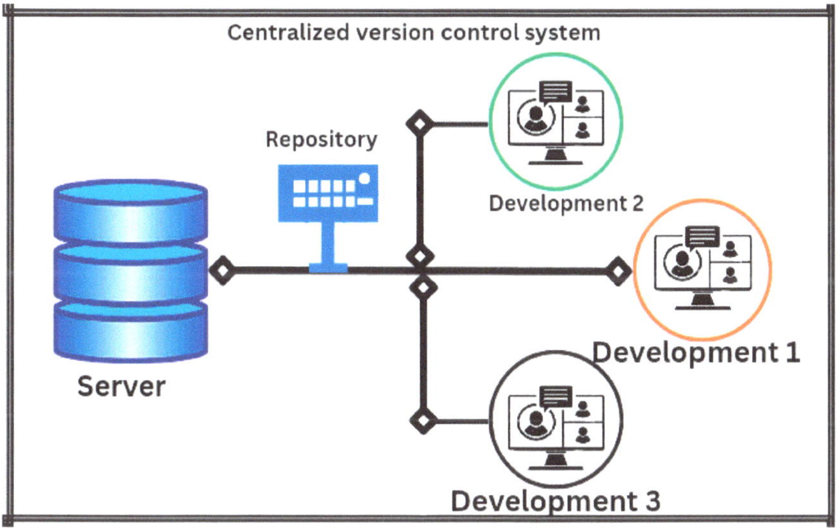

Figure : A view on Centralized Version Control System(CVCS)

The architecture of a centralized version control system (CVCS) is like a client-server model. Where a server acts as the main repository that stores every version of code.

Using centralized source control, every user commits directly to the main branch. It means a file locking is executed. This helps us the checked out code is never accessible to others, ensuring that only one developer can contribute to the code at a time.

Example: SVN(Subversion)

3.1 Introduction to SVN

SVN stands for Subversion. SVN is a centralized version control system. It is an open-source which was developed by CollabNet and came into market in 2000. Subversion has been the biggest winner as an open-source since 2000 until 2016 in the corporate world.
Refer https://subversion.apache.org/

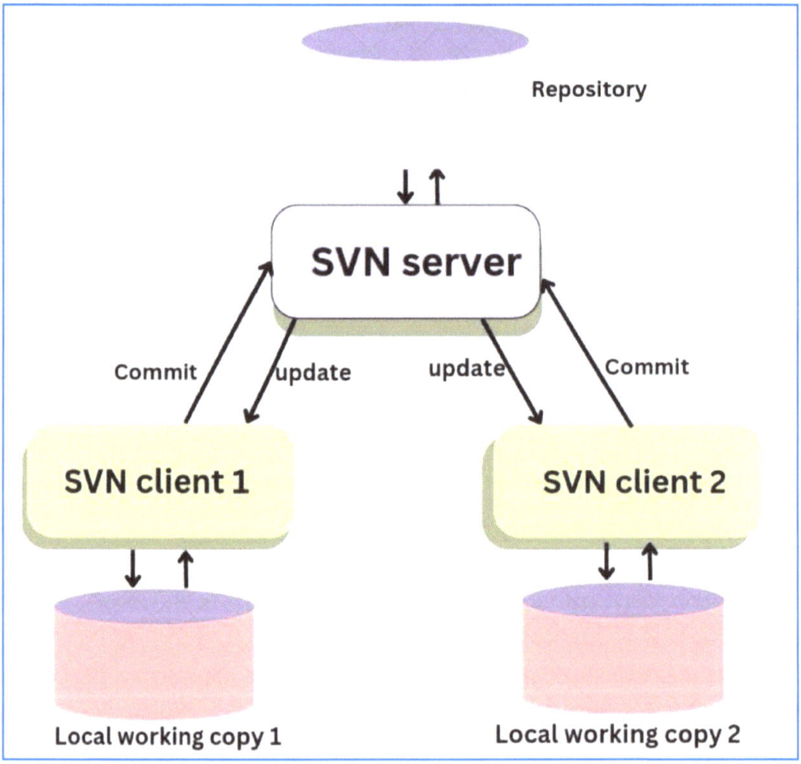

Figure : SVN Basic Architecture

Subversion stores the version history in a central server. Several development teams need to write the code a local computer and change to certain files, they will need to first pull files from that central server to local. Finally, when the changes are done, they need to push their all changed files back to the central server.

Apache Software Foundation developed SVN. It has been a part of a rich community of developers.

Hence the central server plays a crucial role and stores all the data and plays a main role to establish collaboration among team members. Even if It just contains a single repository that comprises several files, each user gets only their working copy.

3.2 History of SVN

CollabNet founded the Subversion project in 2000 which was a pure open-source version-control system. The operations of Subversion were almost like CVS. It was designed to make it better than CVS including some fixed bugs and extra features which were missing in CVS. By 2001, Subversion added a new feature to host its source code. Subversion is developed as a project of the Apache Software Foundation, and as such is part of a wide community of developers and its users.

The Initial release date was October 2000 and was developed in the Programming language: C. The founders of SVN were CollabNet, Apache Software Foundation, Introversion Software, and the FLOSS community.

3.3 Architecture of SVN

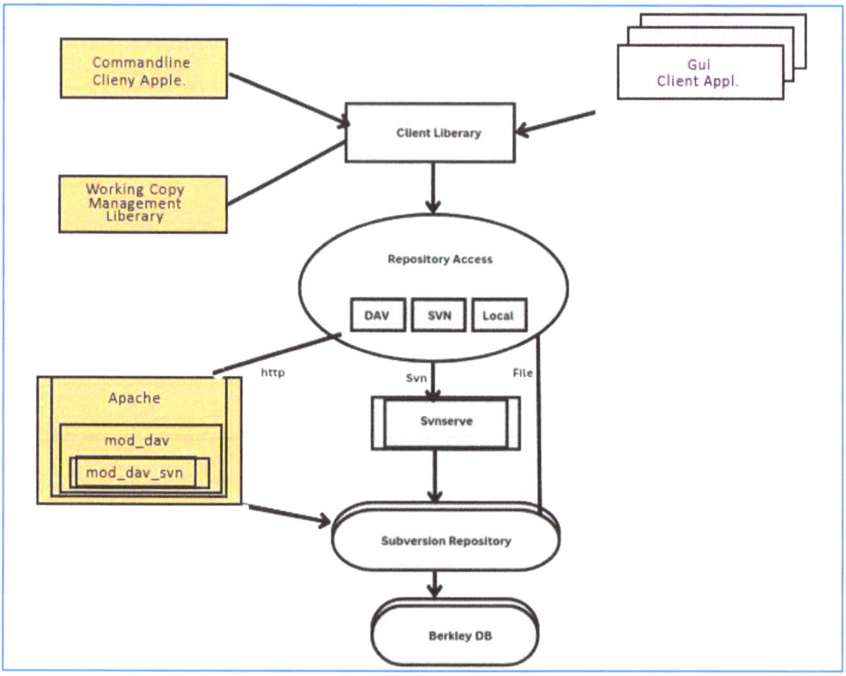

Figure : A deep dive into SVN Architecture

The above figure is architecture of Subversion.

SVN has 2 main elements

- The central server, which keeps project, source files, versions and metadata.
- A local copy of the files, which is on your computer i.e., simple plain working files.

SVN works at 3 layers

Client Layer

The Subversion client is designed on 3 libraries. These are library, libsvn_wc, libsvn_ra, and libsvn_client. These are GUI and command lines.

Here libsn_wc provides an API for managing the client's working copy of a project.

on the other hand, libsvn_ra facilitates with an API to interact with Subversion repository and exchange the information.

Network Layer

A network library (libneon) translates the incoming requests into a set of HTTP requests.

Filesystem Layer

Filesystem library is libsvn_fs. It helps to reach a particular request to a repository which later is interpreted by the Subversion.

Here libsvn_fs interpreted the requests on the arrival at a particular repository.

Developers need to install Subversion client svn at their locally like TortoiseSVN. The client here supports to establish communication with the Subversion repository and processes the working copies provided to it in its local workspace. Several other sub-components play a crucial role to make SVN work through various Repository Access (RA) layers. It means it is conceptually and logically divided into several separable layers.

Subversion is operated via the command line, a corresponding Graphical User Interface(GUI), or via an integrated development environment(IDE).

3.4 Features of SVN

Some of the important features of Subversion are as follows:

- Files and directories are all the time versioned to keep history.
- Atomic commits is inherent feature.
- File conflicts are resolved interactively,
- File locking for the files that cannot be merged.
- Subversion supports unlimited project counts and size.
- Semi-automatic branch and merge tracking,
- Versioning of symbolic links is another feature.
- It supports path-based authorization.
- It has a standalone server option (svn server).
- C#, Perl, Ruby, PHP, Python, and Java are bind with SVN.
- Efficient delta creation algorithm, only changes are stored in the process,
- Tagging - the naming of file versions,
- A strong syntax for external definitions
- Free-form versioned metadata ("properties")
- Costs are proportional to change in the size, not data size
- SASL authentication support for the svnserve server.
- Directories are versioned
- Apache network server option, with WebDAV/DeltaV protocol

3.5 SVN Terminologies

svn
The command-line client program.

trunk
It contains the latest source code, which is in development

tags
It contains a snapshot of the project

branches
It contains different branches of a project

svn version
It reports the state of a working copy in terms of revision.

svnlook
It is used for inspecting a Subversion repository.

svnadmin
A tool for creating, tweaking, or repairing a Subversion repository.

svndumpfilter
It is a program used for filtering Subversion repository dump streams.

mod_dav_svn

Apache HTTP Server use this tool as a plug-in module. It is used to make your repository available to others over a network.

svnserve

It is a daemon process .It is invokable by SSH. It is a custom standalone server program, runnable in background.

svnsync

A program for incrementally mirroring one repository to another over a network.

3.6 Installing SVN

The Apache Subversion project does not support or maintain any binary packages of the Subversion software.

For reference follow below URLs:
https://subversion.apache.org/download.cgi#recommended-release
https://subversion.apache.org/packages.html/

There are several SVN server software (Client and Server) available for several kinds of operating systems. A few of them are listed below:

To install Subversion from here
https://subversion.apache.org/

You can also just use any SVN tool to just create local repositories on your file system if you like.

Keep in mind that to set up VisualSVN, you need to download and install the latest stable version of the TortoiseSVN Subversion client.

For the client side, you can use Tortoise SVN from here https://tortoisesvn.net/downloads.html or Visual SVN from here
https://www.visualsvn.com/downloads/

Let's try to Download and install the VisualSVN server
https://www.visualsvn.com/server/download/

System Requirements.

Operating Systems	Windows Server 2012 R2 – 2022 or later
Minimum Hardware	1.4 GHz CPU, 512 MB
Compatibility	Built against RAM the latest Subversion 1.14 LTS release.
Client	Compatible with all Apache Subversion clients.
Hard Disk	50 MB hard drive space

Table : A view on System requirements for SVN Client

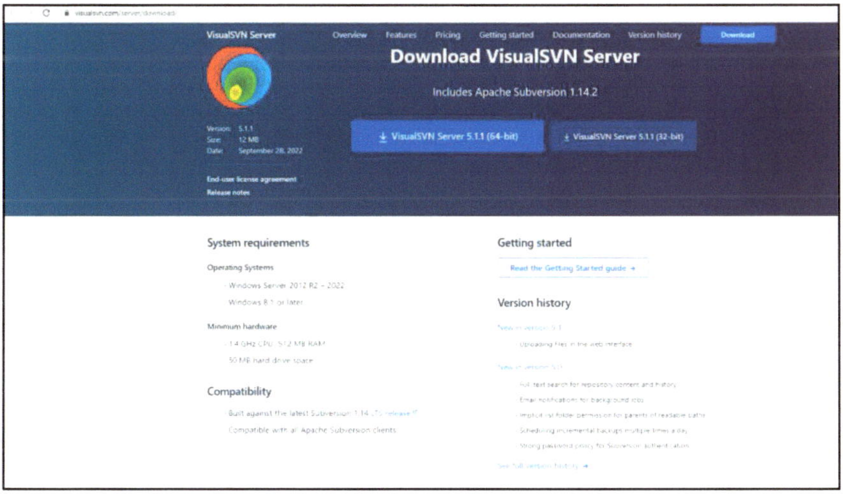

Figure: Downloading SVN client

Accept the License Agreement
Review and accept the End User License Agreement (EULA) https://www.visualsvn.com/server/licensing/eula

Figure : A view on License Agreement

Choose option to install VisualSVN Server and Administration Tools components when deploying VisualSVN Server.

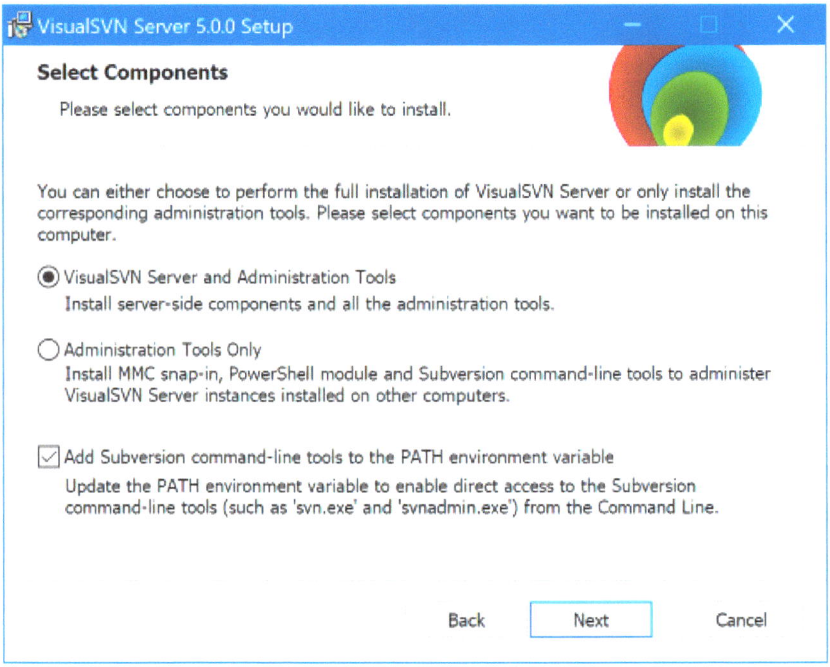

Figure : Selection of Admin Tool options

Adjust the initial server configuration before starting the installation.

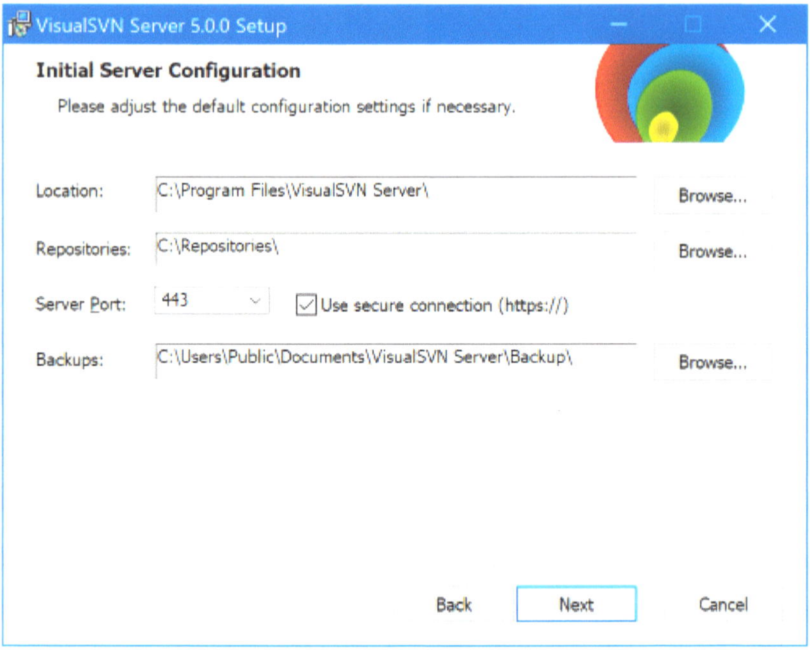

Figure: Initial server configuration

Location: It is a installation directory.

Repositories: It is a root directory where all the Subversion repositories will be installed.

Server Port:
It requires a TCP port that will be used by VisualSVN Server for client access to the repositories over HTTP(S). Server port can BE changed after installing the server.

Use secure connection: It is to use secure HTTPS protocol for client access to the repositories.

Backups. Specifies the default destination location to be used for Subversion repository backups.

Configure search indexing for repositories

Consider enabling the full-text search feature that allows finding files with the specified content in the web interface. To enable the full-text search capabilities, you have to enable search indexing for repositories.

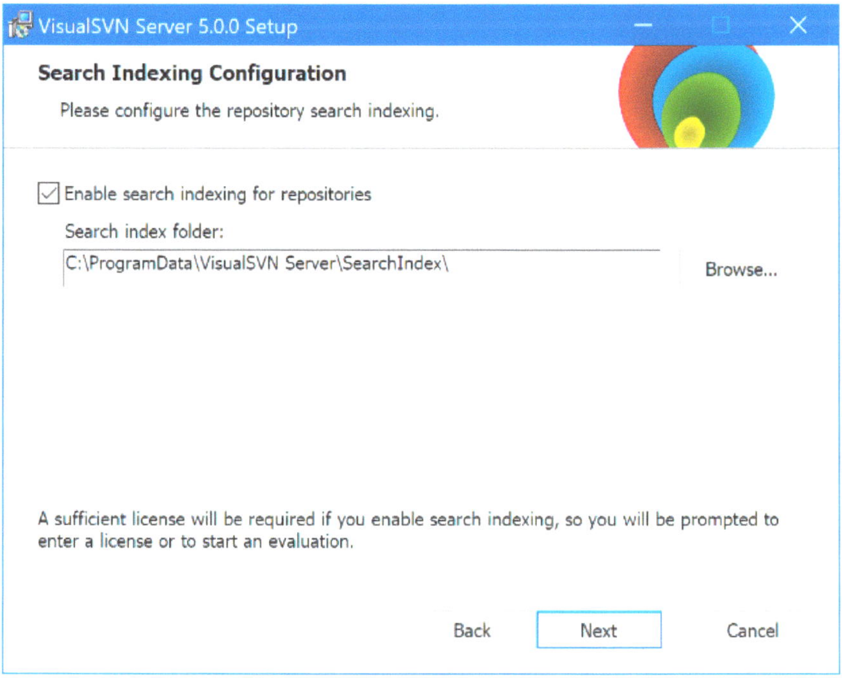

Figure : Search Indexing configuration option

If you enable search indexing, it is generally recommended to change the default Search index folder location. Alternatively, you can keep the search indexing disabled during the first install and configure it later via the VisualSVN Server Manager console.

Configure the SSL/TLS certificate

Actually installer will automatically generate a self-signed certificate if there is no Active Directory domain, the If the computer is joined to an Active Directory domain, the installer will present two options: either to Create a self-signed certificate or to Obtain the certificate from Active Directory.

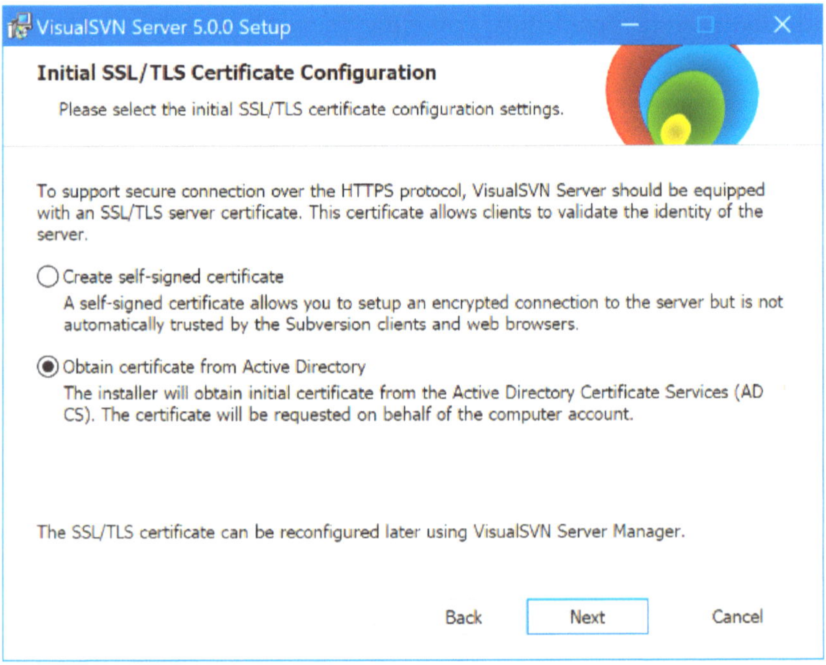

Figure : SSL/TLS configuration

Select the authentication mode

There are 2 authentication modes are available : Subversion authentication and Windows authentication. Later is for Active Directory.

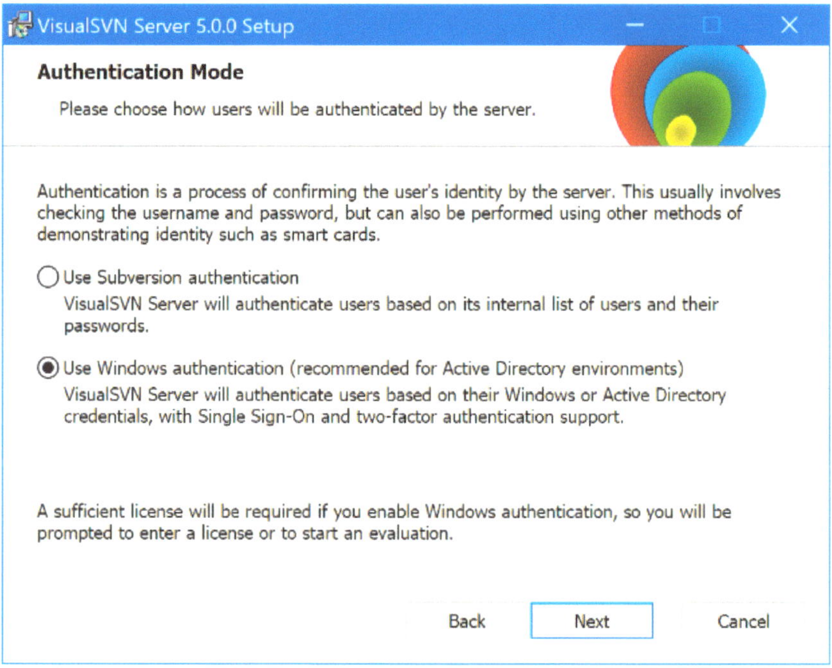

Figure : Authentication mode

Start a 45-days evaluation or enter a sufficient license key

VisualSVN Server is available under the following main license types: Community (free), Essential (paid), and Enterprise (paid). When you choose to enable repository search indexing or Windows authentication, you need to start a 45-day evaluation or enter a sufficient license key because these features are not available under a free Community license.

Select a suitable option, provide the license key if needed, and click Next.

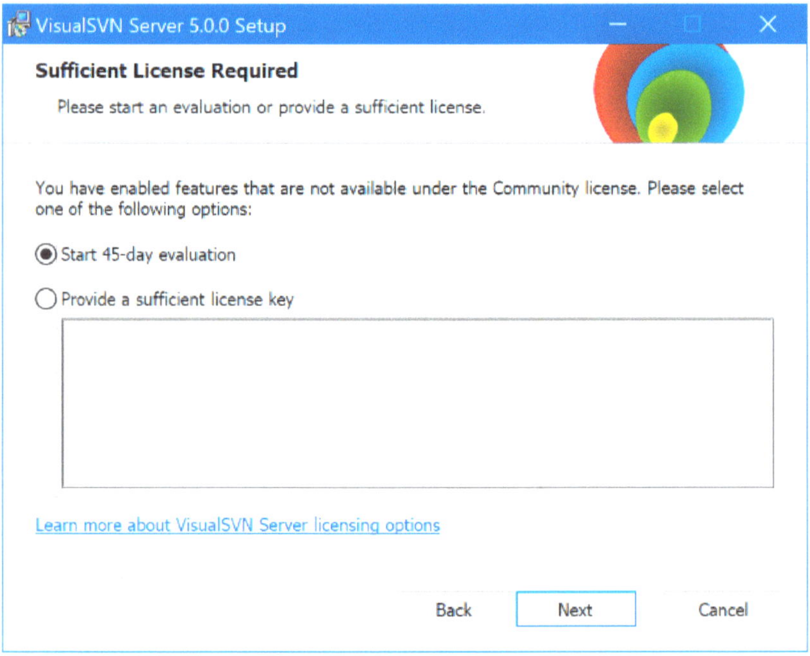

Figure: License requirements key

Finish the installation

Click Finish to close the installer and proceed to the server configuration steps with the VisualSVN Server Manager console.

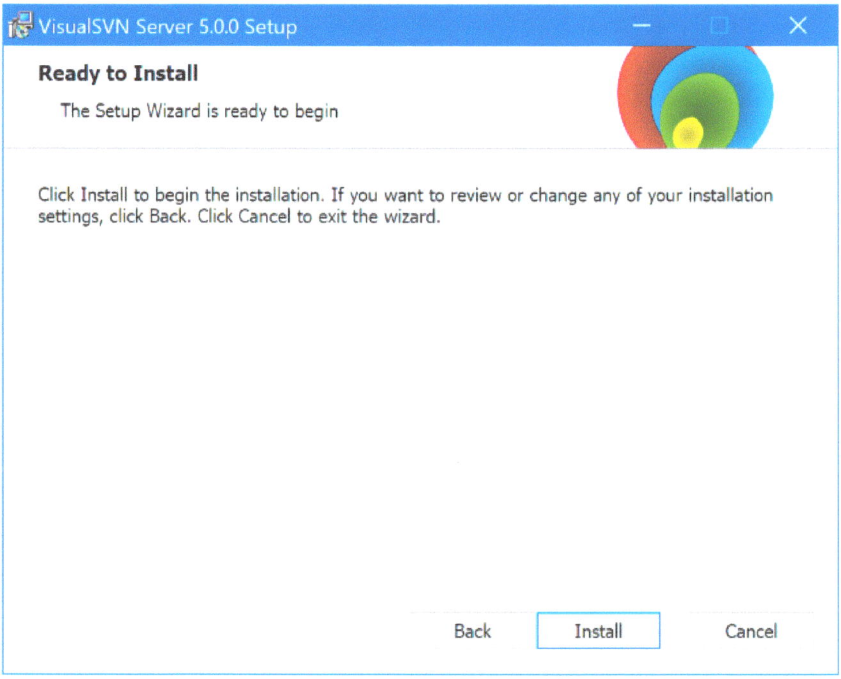

Figure: SVN Client is ready to be installed

Troubleshooting

VisualSVN Server configures and starts VisualSVN Server service that makes Subversion repositories available via the HTTP(S) protocol. If the service does not start during the installation read this article https://www.visualsvn.com/support/topic/00029/

VisualSVN Server Manager is now ready for use

The moment the installation is completed, VisualSVN Server Manager will launch. The console provides an GUI for server and repository administration.

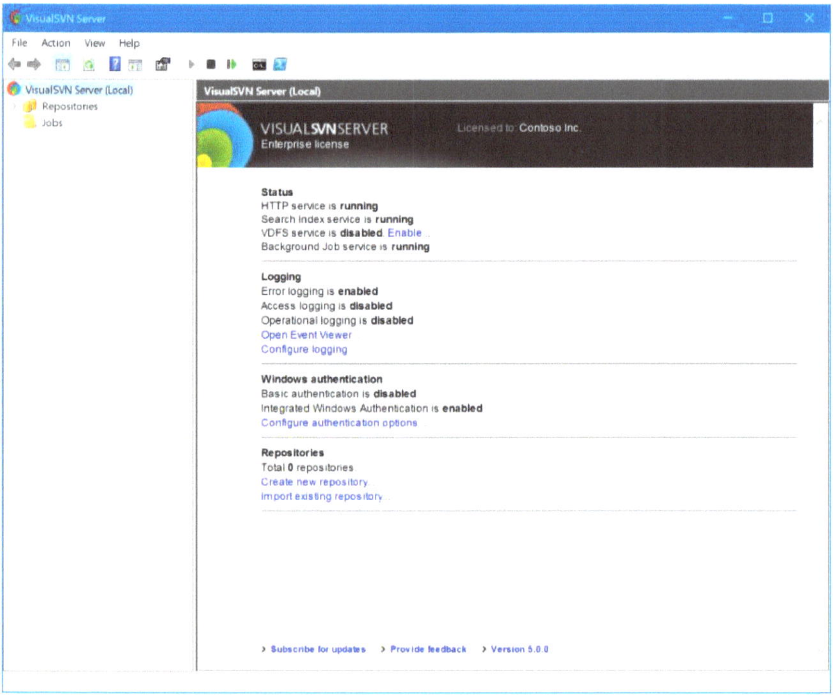

Figure: A Final view after Installation

Performing Initial Configuration

Now we need the initial server configuration

Configuring Network Settings

Now you need to configure network settings.

Follow these steps to open up the network settings:

- Start the VisualSVN Server Manager console.
- Click Action | Properties.
- Click the Network tab.

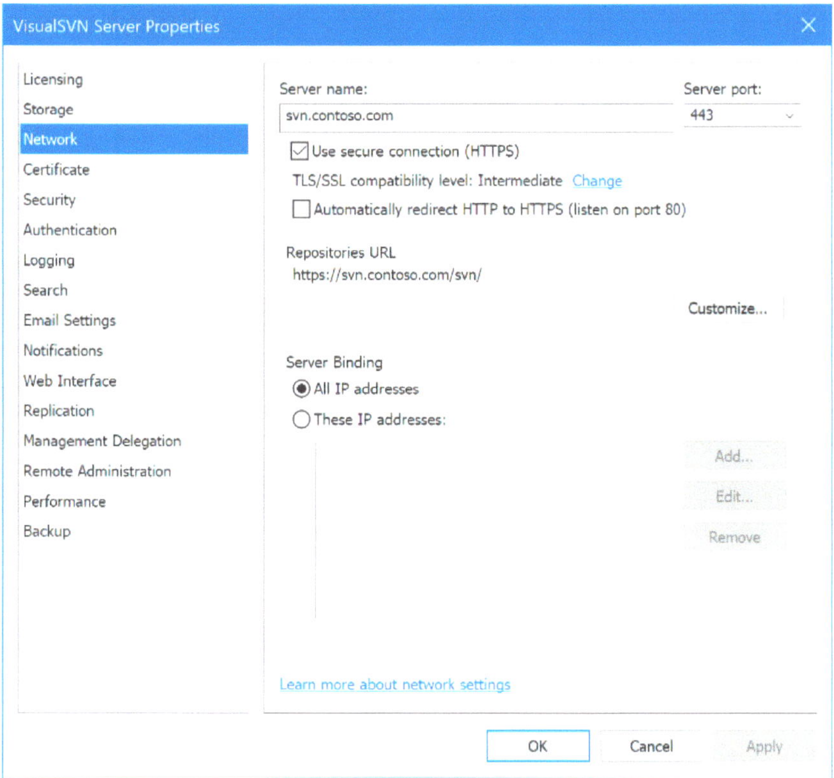

Figure: Fill the required Inputs during setup

Server name:
By default, a local server hostname is considered as the server name. Later you can manually adjust the setting to include a new DNS name.

Automatically redirect HTTP to HTTPS (listen on port 80):
VisualSVN Server can help you transparently switch from plain HTTP to secure HTTPS protocol.

Consider enabling this setting if you plan to switch access to the repositories from HTTP to HTTPS.

Repositories URL

All the repositories hosted on VisualSVN Server are nested under a single repository root in the URL. By default, the root URL prefix is svn/. You can adjust the root URL prefix if required.

Server Binding

VisualSVN Server binds to all available IP interfaces, by default.

Configuring User Authentication Settings

You should always review user authentication settings before starting to add the Subversion repositories to a newly installed VisualSVN Server instance.

VisualSVN Server supports two distinct authentication modes:

Subversion authentication is suitable for non-domain VisualSVN Server installations where Windows authentication is suitable for VisualSVN Server installations in Active Directory.

Configuring SSL Certificate

For secured connection, VisualSVN Server must be installed with a valid SSL certificate.

If the certificate was obtained from the Active Directory and you didn't change the Server name on the Network tab, So, no additional actions are required.

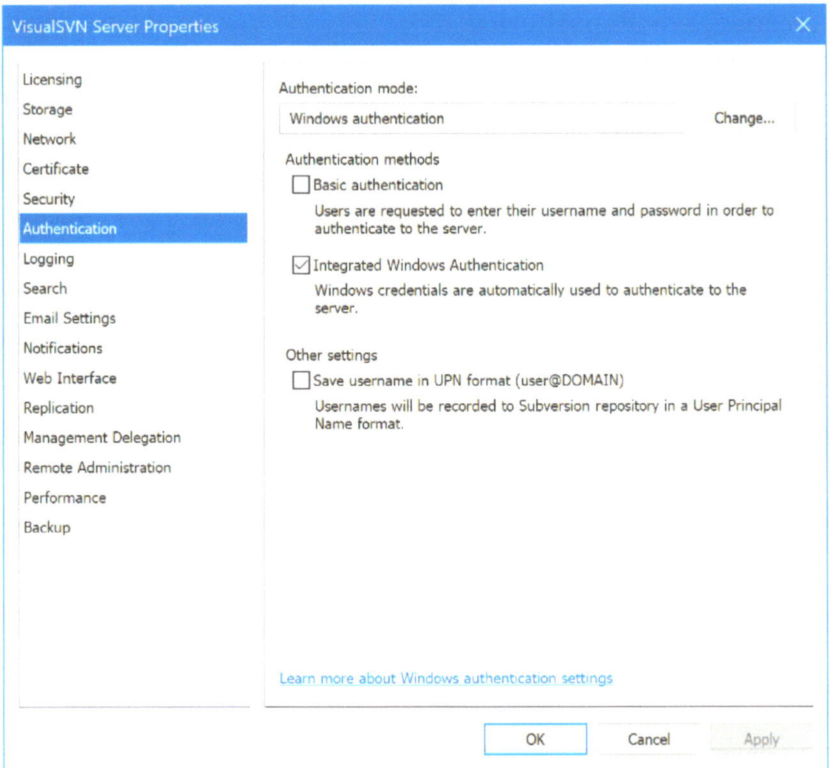

Figure: Authentication mode in SVN client

Importing Existing Repositories:

After Installation and staring it first time VisualSVN Server instance will be empty and does not include any default repositories. You must create new or import existing Subversion repositories into

VisualSVN Server. VisualSVN Server Manager console helps you add the repositories in a straightforward wizard-driven mode.

To import an existing repository, run the Import Existing Repository wizard in VisualSVN Server Manager.

Start the VisualSVN Server Manager console.
Right-click the Repositories node.
Click Import Existing Repository.
Follow the steps of the wizard to complete the task.

Creating New Repositories

To create a new repository, run Create New Repository wizard in VisualSVN Server Manager:

Start the VisualSVN Server Manager console.
Right-click the Repositories node.
Click Create New Repository.
Follow the steps of the wizard to complete the task.

Configuring User Permissions

You need to add access permissions when you create new or import existing Subversion repositories.

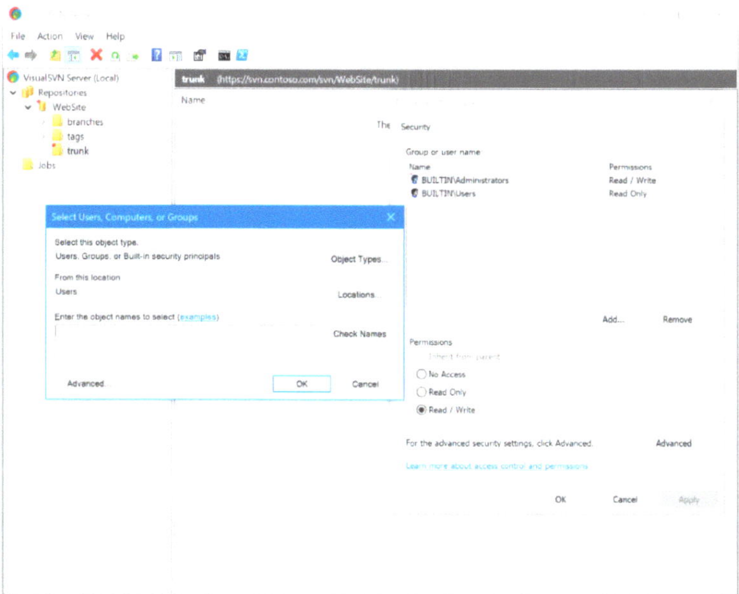

Figure: Configure user permissions

VisualSVN Server supports Read / Write, Read Only, and No Access levels. User permissions on any level including per-folder and per-file levels needs to be managed.

VisualSVN Server Manager console provides an intuitive graphical interface for access control. You can define and manage user access permissions without having to deal with command line tools and text-based configuration files.

Configuring antivirus exclusions

During the installation, VisualSVN Server automatically adds necessary antivirus exclusions to the built-in Windows Defender antivirus.

Final Test and Login

VisualSVN Server makes Subversion repositories available to users through the HTTP(S) protocol. A web browser or Subversion client including command-line svn.exe and GUI clients is used to access the data.

3.7 Using TortoiseSVN Client

You can download the TortoiseSVN Client at
https://tortoisesvn.net/downloads.html
To access the repository, you should first obtain its URL. You can always find the correct repository URL using the VisualSVN Server Manager console:

- Start the VisualSVN Server Manager console.
- Expand the Repositories node.
- Right-click a repository or repository path and click Copy URL to Clipboard.
- You can paste the URL into the URL field of your Subversion client.

Example:
svn checkout https://svn.contoso.com/svn/MyProject/trunk my working project

It appears in the familiar right-click (context-click) menu. It only takes a few steps to get started.

Right-click on a file or folder.
Then choose TortoiseSVN among choices like Open, Cut, Copy, etc.

Creating a Repository using TortoiseSVN

First, create a new empty directory on your PC.
\svn_repos. Now right-click on the new folder and from the context menu choose TortoiseSVN → Create
Repository here… The repository is then created inside the folder, ready for you to use.

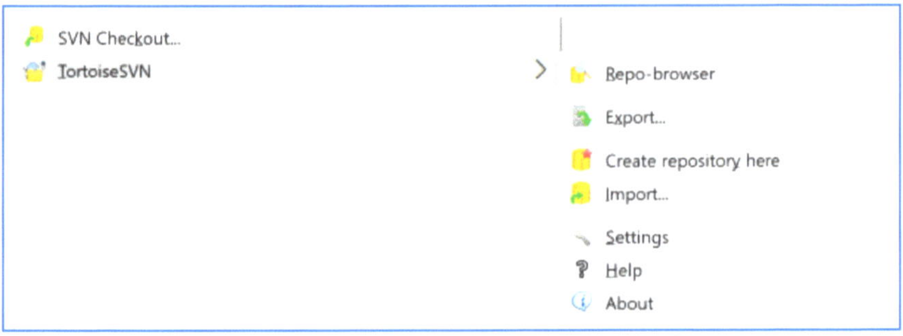

Figure : Create repository in SVN client

Using TortoiseSVN

Importing a Project using TortoiseSVN

We created a repository, but it is empty. Let's assume I have a set of files

\Projects\ myproject that I would like to add. Navigate to the myproject folder in Explorer and right-click on it.
Now select TortoiseSVN → Import.

Checking out a Working Copy using TortoiseSVN

After we have a project in our repository, we need to create a working copy

Now we will need to check out the project folder of our repository into a development folder on the PC called C:\Projects\myproject-Dev. Create that folder, then right-click
on it and select TortoiseSVN → Checkout… Then enter the URL to checkout, in this case,
file:///c:/svn_repos/trunk/myproject and click on OK.

Our development folder is then populated with files from the repository.

Viewing the Project History using TortoiseSVN
One of the most useful features of TortoiseSVN is the Log dialog.

Undoing changes using TortoiseSVN

If you want to remove the changes you have not committed yet and reset your file to the way it was before you started editing,
Use this TortoiseSVN → Revert

Server-side hook scripts using TortoiseSVN

A hook script is a special type of program that can be triggered by some repository event. These events could be creation of a new revision or the modification of an unversioned property.
The hook can be a batch file or an executable. Some default hook scripts can be found in the hooks directory of the repository.

Example
rem Only allow log messages to be changed.
if "%4" == "svn:log" exit 0
echo Property '%4' cannot be changed >&2
exit 1

Importing Data Into a repository using TortoiseSVN

Select the top-level folder of your project directory structure in windows explorer and right-click then open
the context menu.
Select the command TortoiseSVN → Import

Before you import your project into a repository you should:
remove all files which are not needed to build the project (temporary files, files generated by a compiler e.g., *.obj, compiled binaries, ...) and proactively organize the files in folders and sub-folders.

Committing your changes to the repository using TortoiseSVN

Sending the changes, you made to your working copy is known as committing the changes. But before you commit
you to have to make sure that your working copy is up to date.

You can either use TortoiseSVN → Update directly.
Or you can use TortoiseSVN → Check for Modifications first, to see which files have changed locally or on the server.

Update your working copy with changes from others using TortoiseSVN

You should always check that changes done by others get incorporated into your local working copy

To update, select the files and/or directories you want, right-click, and select TortoiseSVN → Update in the explorer menu.
Changes done by others will be merged into your files, keeping any changes you may have done to the same files.

These are color coding to highlight different update actions:

Green: Changes from the repository successfully merged with your local changes.
Purple: New item added to your WC.
Black: Unchanged item in your WC updated with the newer version from the repository.
Dark red: Redundant item deleted from your WC, or the missing item replaced in your WC.
Light red: Changes from the repository merged with local changes, resulting in conflicts that you need to resolve.

3.8 Basic SVN Commands

Here are the basic SVN commands that every developer and admin should know.

svn admincreate
It creates a new, empty repository.

svn import
It commits an unversioned tree of files into a repository

svn checkout
It checks out a working copy from the repository. This command is sometimes shortened to svn co.

svn commit
It sends your changes back to the SVN server.

svn add
It add a new file to the repository only after svn commit.

svn delete
It delete a file from your working copy of the repository.

svn list
It shows a list of files in a repository without creating a working copy.

svn diff
it display the differences between your working copy and the copy in the master SVN repository.

svn status
It tells the status of working copy files and directories.

svn info
It displays information about a local or remote item.

svn log
It dispaly log messages from the repository.

svn move
It either moves a file from one directory to another or renames the existing one.

svn merge
It merge two different versions into your working copy.

svn revert

It reverts changes in your working copy, as well as property changes.

svn update
It update command updates your working copy with changes from the repository.

svn shelve
It save the changes without submitting them.

svn help
It provides any help a developer seek for a summary of available commands.

Chapter 4
Distributed Version Control System

A distributed version control system (DVCS) is a type of version control local copy of the repository is pulled from the main server, we can work locally and offline. A developer can retrieve any version of a specific file as it has a local history that can be used to commit our changes whenever needed. It means that our local history can diverge from the remote one and the action of pushing back our changes to the main server is equal to a merge of both histories.

Distributed VCS introduced significant advantages over Central VCS. In Distributed VCS, a local copy of the project is resided at developer's machine along with its history. Distributed VCS architecturally different from a client-server approach of Central VCS to a peer-to-peer approach for version control. That is, in the case of Central VCS the central server helps to keep history at remote with the latest code from the local machines of the developers.

Some examples of DVCS are **Git and Mercurial**.

4.1 Introduction to GIT

Linus Torvalds in 2005 introduced GIT. As we know that older centralized version control systems such as SVN and CVS has some issues, Git is distributed and solve several problems. Here every developer keeps the project i.e. repository the full history of their code repository locally which seems that initial clone of the repository and its operation slower at first time but subsequent operations such as commit, blame, diff, merge and log significantly faster.

Git architecture supports a great branching & merging strategy and rewriting repository history with the help of powerful sophisticated tools. Staging area , hooks and pull requests are one such popular tool that allows teams to collaborate. Several Git branches can be used to merge each other and reviewer can review other's code. Today, Git is widely accepted and adopted version control system in the software field and is considered the modern standard for software development.

Git has these below advantages

- The ability to create branches with ease
- The ability to add one or more remote repositories and synchronize with it
- Local repository with it owns local history
- Low space consumption
- A useful tool that helps to find which commit is responsible for a regression (this will be covered in another blog post)
- Great performance

4.2 Architecture of GIT

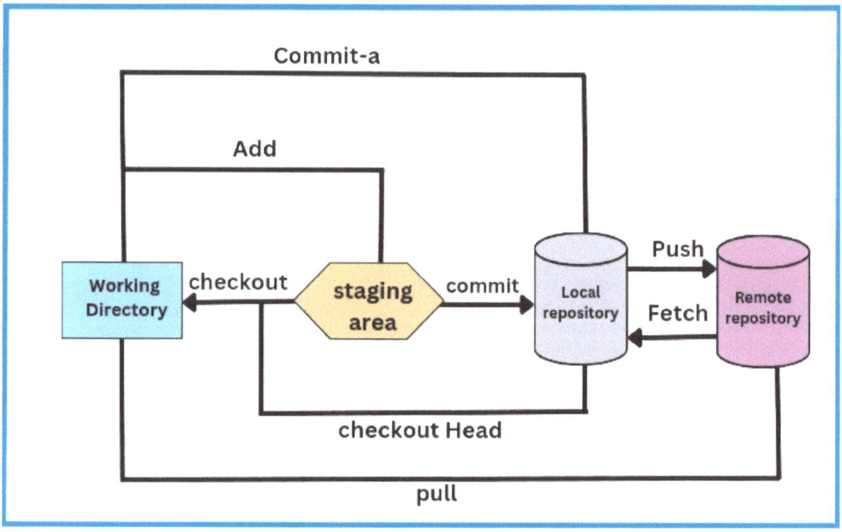

Figure: Architecture of GIT

Git uses a 3-layered architecture. The 3 layers are:

Working Directory
This is created onto your local machine and allows you to edit the source code available at your local. when a Git project is cloned or initialized using git clone or git init commands.

Staging Area
This a special area where the code is staged and stayed until it is pushed by applying the command, git push. Git add command is used to add the code into staging area. In case further modifications are made in the working directory, the snapshots for these two layers will be different. However, these can be synced by using the same 'git add' command.

Remote Repository
A remote repository or project is a repository at remote server.

Types of branches in GIT

There are five different branch types in total:

- Master Branch
- Develop/Integration Branch
- Feature Branch
- Release Branch
- Bugfix Branch
- Hotfix Branch

Master Branch

This is the main branch which keeps the latest stable code of production. A full restriction policy must be applied to access to Master branch and direct merge. It should always have the latest stable version of the production server. Allow access to only CD tools like Jenkins to make commits to this branch.

Develop/Integration branch

This is the most important and active branch of the repository from which we make releases to the production server. Code is merged from this branch to release branch when it becomes eligible for production deployment. The code in this branch should always be in a deployable state to production.

Feature Branch

For every feature added, there should be a feature branch. This ensures parallel developments making rollbacks and cherry-picking easier. The feature branch must be branched out from Develop branch.

Bugfix Branch

If the code changes made from the feature branch were rejected after a release, sprint, or demo, any necessary fixes after that should be done on the bugfix branch.

Hotfix Branch
If there is a blocker and require urgent temporary patch, or apply a critical configuration change that should be handled immediately, it should be created as a Hotfix.

4.3 Features of GIT

Distributed System
Having a distributed system, Git allows users to work simultaneously on the same project, without interfering with others' work. When a particular user gets done with their part of the code, they push the changes to the repository, and these changes get updated in the local copy of every other remote user which pulls the latest copy of the project.

Speed
Git is very fast and efficient in handling of large projects.
The fetching speed from a local repository is almost 100 times faster than what is possible with the remote server.

Git Submodules
Git provides submodules which are useful when your Git project has dependencies that are Git projects themselves.
Tracks history:
It is designed in such a way that it can handle small and large projects with easy tracking of files.

Reset Branch
Another one of the advanced Git features can be used to reset your branch, revert, or rewind to the last commit. By default, it will rewind to the last commit,

Non-linear Development
A user can pick up any part of the project and do the required operation and then further update the project. This can be done by the Non-linear development behavior of Git.

Staging Area
Unlike the other systems, Git has something called the "staging area" or "index".

Role-Based Codeline
Got provides a role-based authorization and authentication to work on the project and several branches. You could be a Git administrator, developer, reviewer, or editor role.

Lightweight
Since git has a remote repository only keeps the final code to the local repository while cloning is completed, hence only right versions and code is reached to remote. It is lightweight.

Reliable
Git provides a central repository where the data is always backed up in every collaborator's local repository. Hence, in the event of crashing the central server, the data will never be lost as it can be gained back easily by any of the developer's local machines.

Simplicity
Branching in Git is easier and simpler than in other Version Control Systems.

Economical
Git is released under the General Public's License(GPL) and hence it is available for free. Even though some licensed versions are available like Git Bitbucket.

Security
Git uses SHA1 to store all the records in the form of objects in the Hash. Each object collaborates with the use of these Hash keys. SHA1 is a cryptographic algorithm that converts the commit object into a 14-digit Hex code. Hence all project is secured.

4.4 Lifecycle of GIT

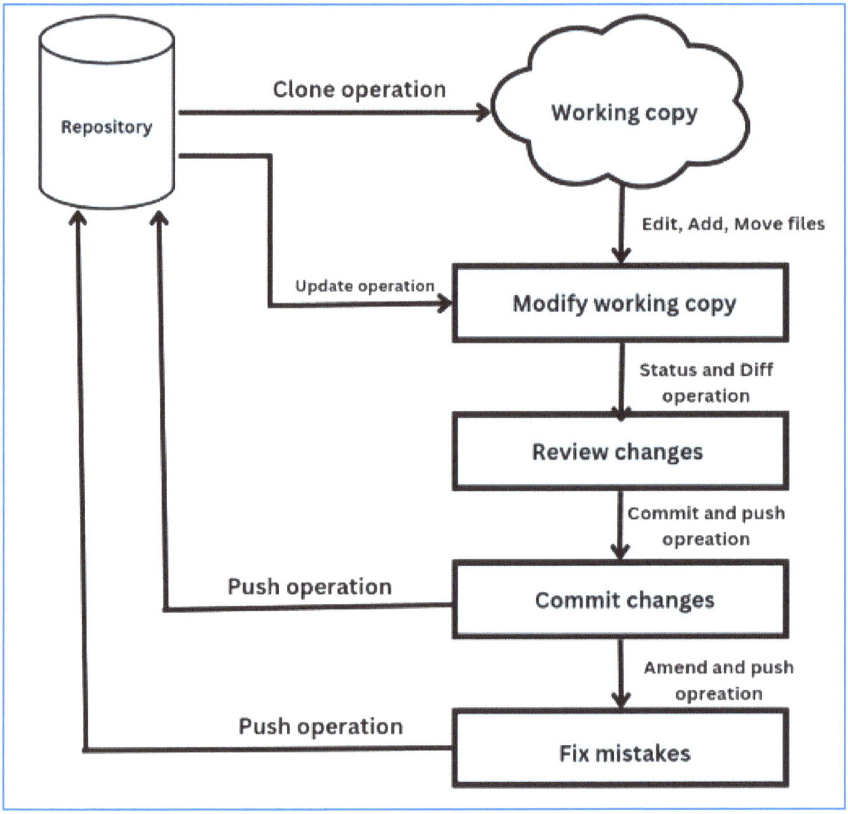

Figure : Lifecycle of GIT

The lifecycle of Git starts at the working directory where the developer commits its local changes. The developer works on several files or creates new files, edits or deletes them, and so on. It is created by cloning a remote project or using the git init command. Next is the staging area. This serves as a holding area to accumulate and stage changes from the working directory before they are committed to the next level—the local repository. Unlike the other systems, Git has something called the "staging area" or "index". This is an intermediate area where commits can be formatted and reviewed before completing the commit. This allows you to stage only portions of a modified file. Gone are the days of making two logically unrelated modifications to a file before you realized that you forgot to commit one of them.

The local environment is built with working directory, staging area, and local repository.

4.5 Types of GIT Objects

A Git object contains your original data files and all the log messages, author information, dates, and other information required to rebuild any version or branch of the project.

There are several subdirectories in the .git folder, and one of them is the .git/objects subdirectory.

It contains 4 different types of objects –

- Blob,
- Tree
- Commit
- Tag.

Blob

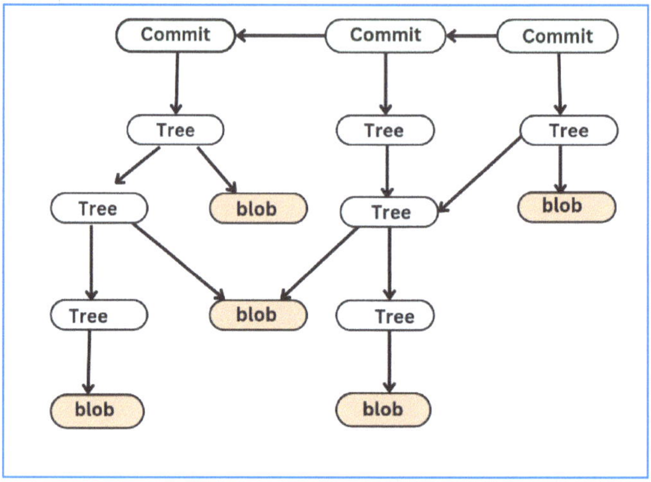

Figure: A blob and Tree

It stands for Binary Large Object. Blob stores the contents of the file. Whenever we commit our files the first type of objects that are created are the blob objects.

Tree Object

Tree objects contain a list of all files in our repository with a pointer to the blob object assigned to them.

Commit Object

Git creates a commit object that has a pointer to its tree object. The commit object contains – a tree object hash, parent commit hash, author, committer, date, and message.

Tag Object

A tag object contains an object name, object type, tag name, the name of the person who created the tag, and a message.

Chapter 5
Installation and Setting up GIT

On your terminal, enter the git --version at the command line to check if you already have Git installed.

Depending on your operating system:

For Windows: Download the Git installer. To open a command window,
go to git bash. vbs from the Git folder of the Programs directory.

For Mac: Download the Git installer. To open a command window, search for the Terminal.

For Linux: Enter *sudo apt-get install git* at the command line. To verify the installation was successful, enter the command
which git
git --version

There are several flavors available for Git installers. In this book, we will consider windows.

The steps for downloading and installing git are as follows.

Browse to any of following
GitHub website https://github.com/ (Server)
Atlassian product https://bitbucket.org/product/download-journey (Server)
GIT SCM https://git-scm.com/ (Client)
Click on the downloaded installer file and select yes to continue.

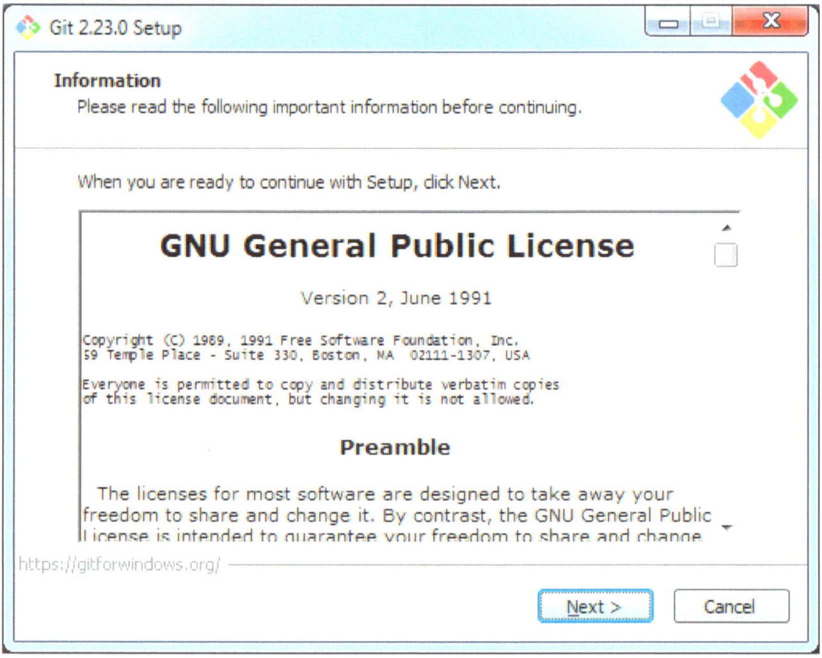

Figure: Git Installer

Default components are automatically selected in this step. You can also choose your required part.

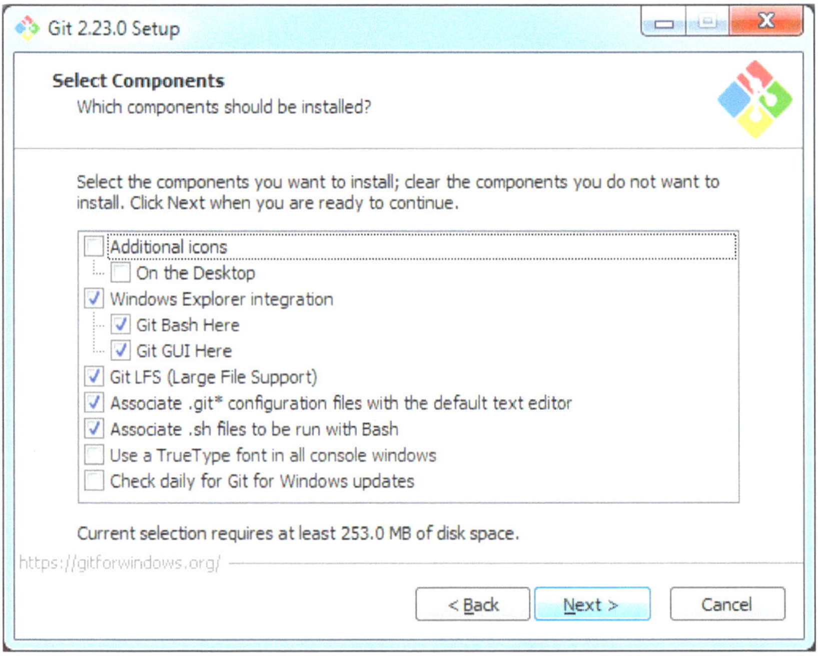

Figure: Select components

If you wish to use a tool from a command line without specifying the full location path, you need to tell Windows about this by adding that directory to the PATH environment variable.

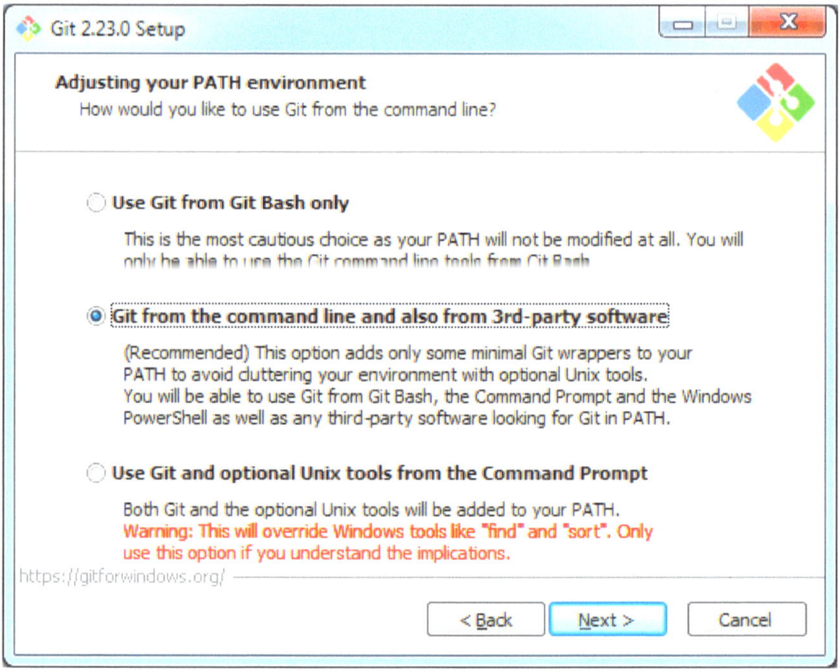

Figure: Adjust PATH environment variables

Unless you know that you have OpenSSH already installed, it is suggested to use the OpenSSH that by default comes with Git.

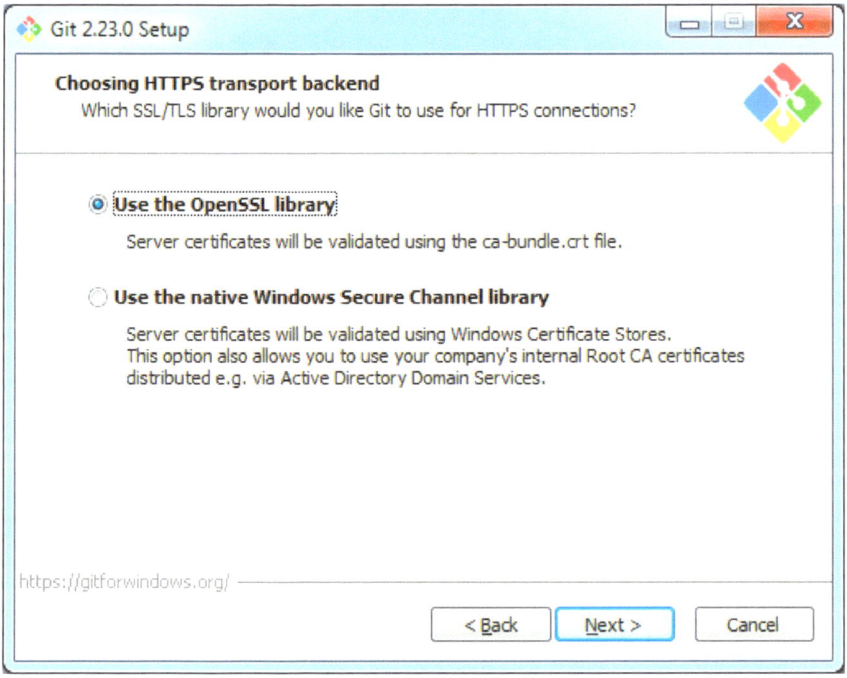

Figure: Choose SSL/TLS options

The next selection converts line endings. It is recommended that you leave the default selection. This relates to the way data is formatted and changing this option may cause problems. Click Next.

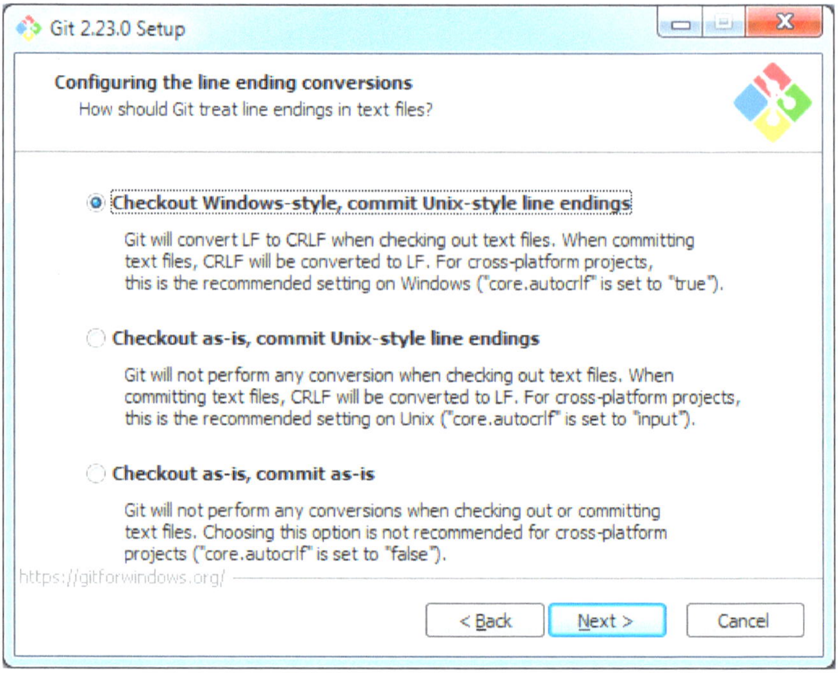

Figure: Choosing different style options for line ending

Choose the terminal emulator you want to use. The default MinTTY is recommended, for its features. Click Next.

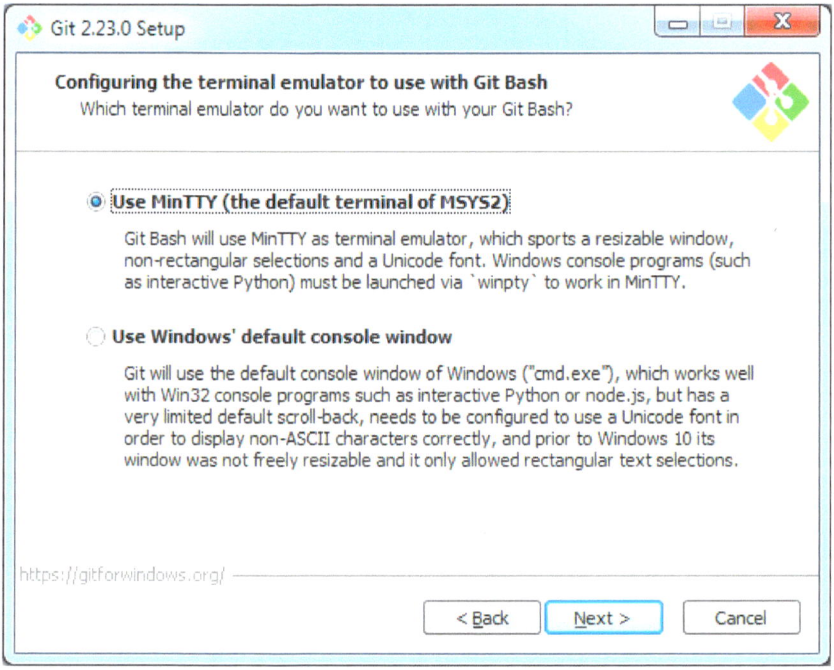

Figure: Choose Terminal options

The default option is recommended unless you specifically need to change its behavior. Click Next to continue with the installation.

Once the installation is complete, you can view the Release Notes or Launch Git Bash, then click Finish.

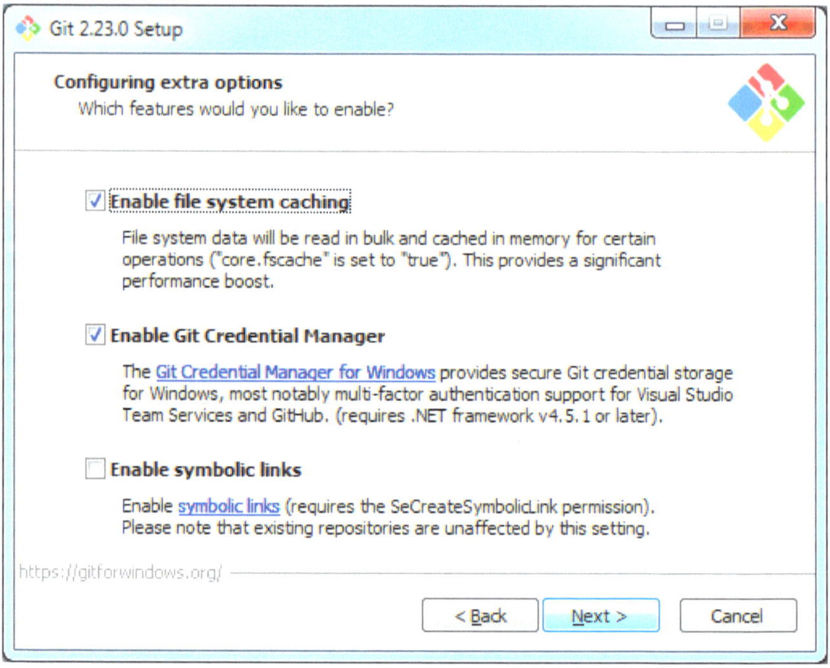

Figure: Choosing the options like caching or credential manager

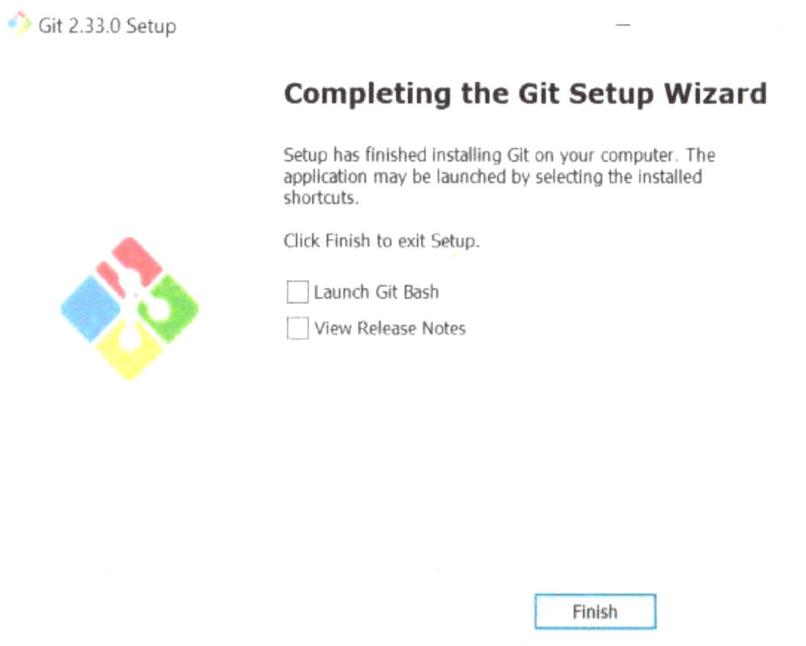

Figure: Git is installed and setup

Open Git bash terminal

Figure: pen Git Bash Terminal at local

You can use the terminal to execute the git command.

5.1 GIT Environment Setup

The Git config command

Git supports a command called git config that is used to set Git configuration values on a global or local project level.

Setting user. name and user .email are the necessary configuration options as your name and email will show up in your commit messages.

Setting username
The username is used by Git for each commit.

$ *git config --global user.the name "Vinay Singh"*

Setting email id
Git uses this email id for each commit.

$ *git config --global user.email "Thedevopsguru@yahoo.com"*

Setting editor

To select a different text editor, such as Vim,

$ *git config --global core.editor vim*

Checking Your Settings

You can check your configuration settings; you can use the git config --list command to list all the settings

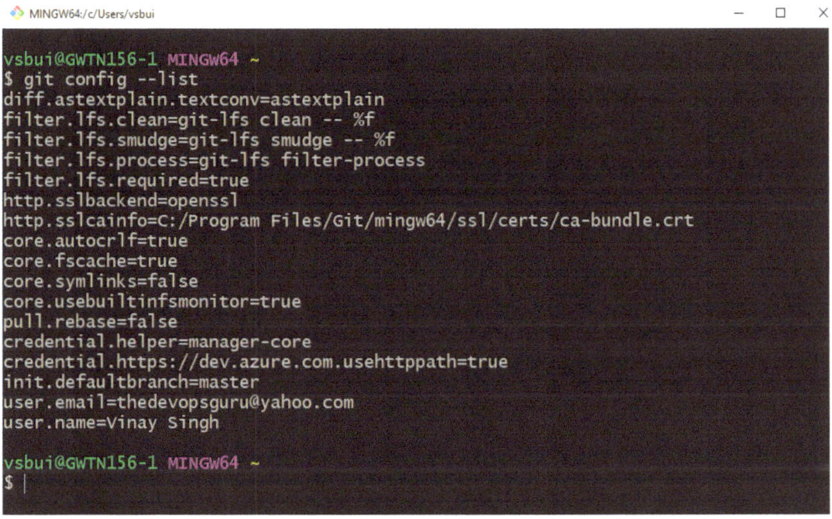

Figure: Git Config list

5.2 Validation of Installation and Setup

To validate the installation, run Git Bash and type

git –help
git --version

5.3 Git Configuration

The git config command is used to specify the configuration level. The following configuration levels are available in the Git config.

global
system
local

It is the default level in Git. Git config will write to a local level if no configuration option is given. Local configuration values are stored in .git/config directory as a file.

--global

This is a user-specific configuration. It's values are stored in a user's home directory. ~ /.gitconfig on UNIX systems and C:\Users\\.gitconfig on windows as a file format.

--system

The system-level configuration is applied across an entire system. The system-level configuration file is stored in the git-config file off the system directory.

Validation of Installation and Setup

To validate the installation, run Git Bash and type

git –help
git –version

Chapter 6
GIT Commands

These are a list of a few commands that you can use frequently on GIT server.

6.1 git branch

List branches (the asterisk denotes the current branch)

git branch -a
List all branches (local and remote)

git branch [branch name]
Create a new branch

git branch -d [branch name]
Deletes a branch

git branch -m [old branch name] [new branch name]
Rename a local branch

6.2 git checkout

it is used to switch between branches.

git checkout -b [branch name] origin/[branch name]
Clone a remote branch and switch to it

git checkout
Switch to the branch last checked out

git checkout -b [branch name]
Create a new branch and switch to it

git checkout -- [file-name.txt]
Discard changes to a file

6.3 git clean

git clean -n

Removes untracked files from the working directory.

Run 'git clean -n' to see a dry run.
Run 'git clean -f' to force untracked file deletion.
Use 'git clean -f -d' to remove untracked directories.
Use 'git clean -f -x' to remove untracked .gitignore files.
use the 'git clean -i switch to do an interactive way.

6.4 git clone

git clone <repo> <directory>
git clone ssh://x_person@example.com/path/to/team-project.git
git clone --branch <tag> <repo>
git clone https://github.com/gchowdhury/git-hello-world.git
It will create a copy of an existing Git repository.

6.5 git commit

Takes the staged snapshot and commits it to the project history. -m option will allow adding add a message as well.

git commit –amend
git commit -m "Your message here for tracking"

Passing the --amend flag to git commit lets you amend the most recent commit.

6.6 git config

A convenient way to set configuration options for your Git installation.

git config --global user.email "your_email@example.com"
git config --global color.ui false

6.7 git fetch

Fetching downloads, a branch from another repository, along with all of its associated commits and files.

git fetch <remote name>
git fetch <remote name> <branch name>
git fetch --all

Some options can be used with git fetch
*--all (will f*etch all remotes).
*--append (-a) (will a*ppends to existing fetched contents without overwriting).
*--no-tags (will d*isable automatic tag).
*--tags (will f*etch all tags from a remote)
*--set-upstream (will a*dd upstream tracking)

6.8 git init

Initializes a new Git repository. If you want to place a project under revision control.

git init
git init <directory>

6.9 git log

View changes for a project

git log --summary
View changes (detailed)

git log --graph
Show logs in a graphical way

6.10 git merge

Merge a branch into the active branch

git merge [source branch] [target branch]
Merge a branch into a target branch

6.11 git pull

Update the local repository to the newest commit

git pull origin [branch name]
Pull changes from the remote repository from a specified branch

6.12 git push

It lets you update your local changes to the remote branch.

git push origin --delete [branch name]
Delete a remote branch

git push origin [branch name]
Push a branch to your remote repository

git push -u origin [branch name]
Push changes to the remote repository (and remember the branch)

git push origin --delete [branch name]
Delete a remote branch

6.13 git stash

Stash changes in a dirty working directory

git stash clear
Remove all stashed entries

6.14 git rebase

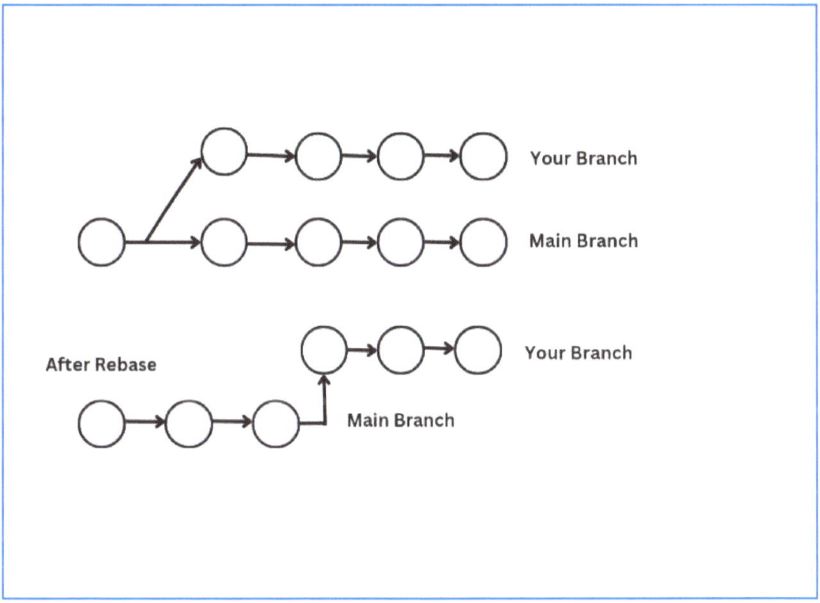

Figure: Git rebase command

Rebasing lets you move branches around, which helps you avoid unnecessary merge commits.
The -i flag is used to begin an interactive rebasing session.

git rebase <base>
git rebase --interactive <base>

6.15 git reflog

Git keeps track of updates to the tip of branches using a mechanism called reflog.

git reflog
git reflog show HEAD
git reflog show [branch name
git reflog show –all

6.16 git remote

The git remote command lets you create, view, and delete connections to other repositories.

git remote add origin ssh://git@github.com/[username]/[repository-name].git
Add a remote repository

git remote show origin
View your current remote repository

git remote set-url origin ssh://git@github.com/[username]/[repository-name].git
Set a repository's origin branch to SSH

6.17 git diff

git diff [source branch] [target branch]
Preview changes before merging

6.18 git reset

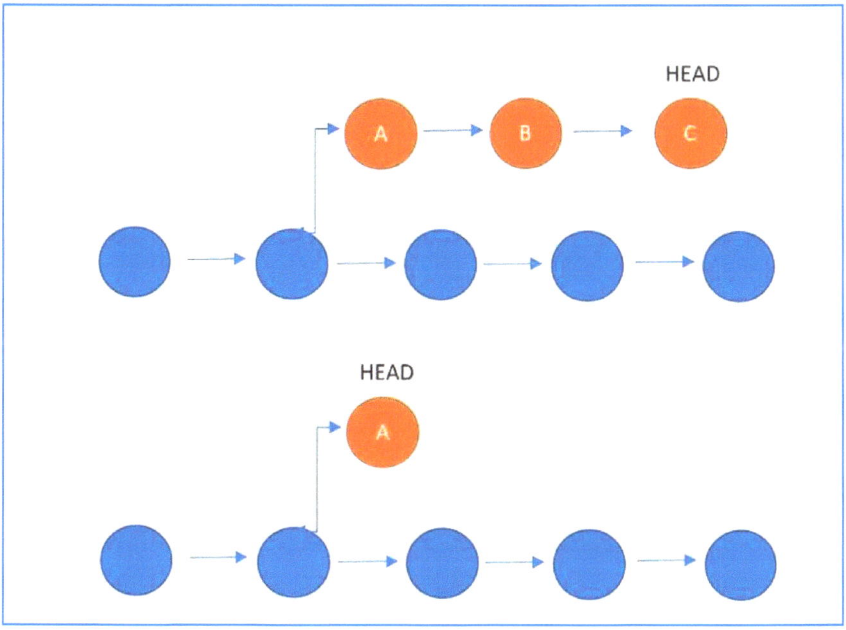

Figure: Git reset command

Undoes changes to files in the working directory. Resetting lets you clean up or completely remove changes that have not been pushed to a public repository.

git reset --hard
git reset --mixed
git reset --soft
git reset --soft 36374547

6.19 git revert

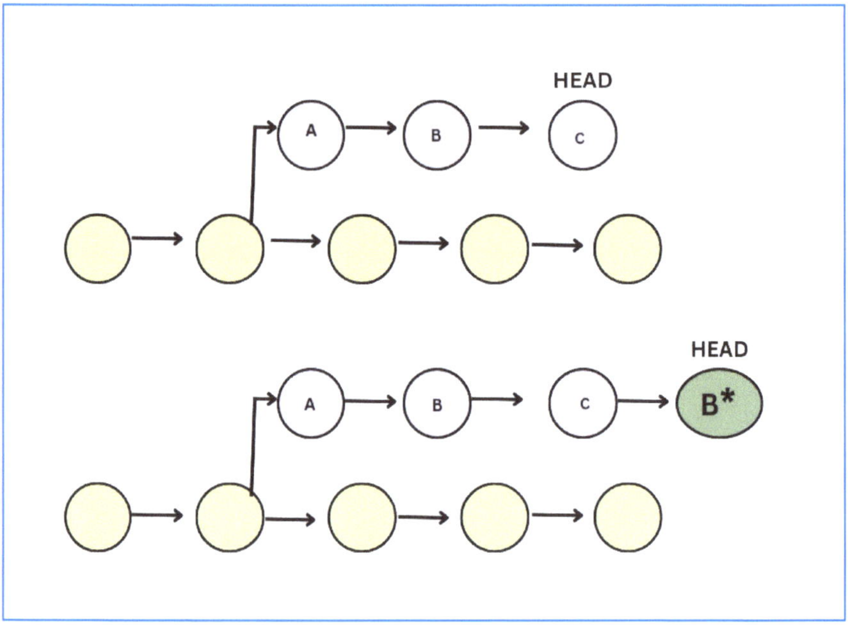

Figure: Git revert command

The git revert command is a forward-moving undo operation that offers a safe method of undoing changes.

git revert HEAD
git revert --edit 98cdh27
git revert --no-edit 98cdh27
git revert --no-commit 98cdh27

6.20 git status

Displays the state of the working directory and the staged snapshot.

git status

6.21 git cherrypick

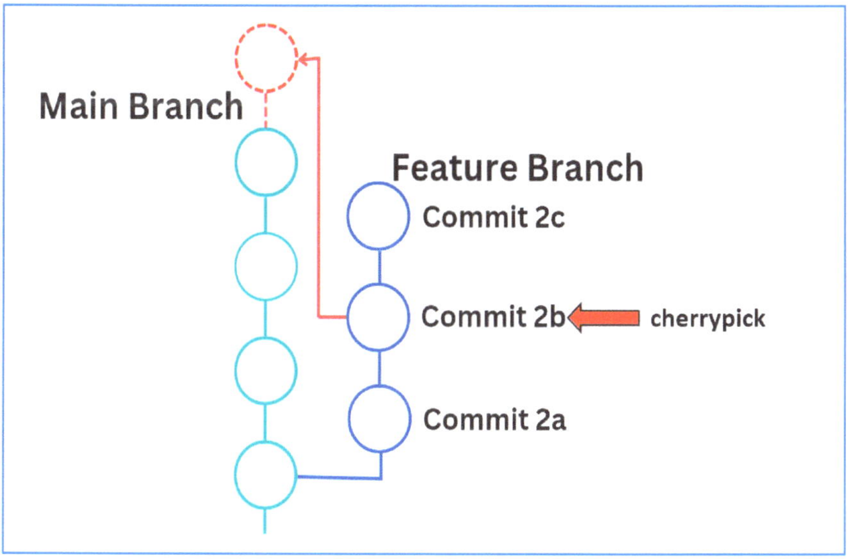

Figure: Git cherrypick command

git cherrypick commit hash

To deploy selected features

git cherry-pick [–edit] [-n] [-m parent-number] [-x] <commit>
git cherry-pick commitSha
git cherry-pick -x <commit-hash>
git cherry-pick f
git notes copy <original-commit-hash> <new-commit-hash>
git cherry-pick A^..B

6.22 git version

git --version
Show git version

6.23 git add

git add filename. extension

It will add the file to the staging directory

git add <file>
git add hello.py

6.24 git rm

git rm filename. extension
Delete a file from Git AND local

git rm -r directory_name
Delete a directory from Git and local

git rm -r --cached directory_name
Delete a directory from Git but not LOCAL

git rm -f filename.extension
Forcefully Delete a staged file

6.25 git shortlog

git shortlog -sn --all --no-merges
To view contributions by each member of your team,

6.26 git tag

Git tag is used for a marked version release.

git tag -a "version"

6.27 git tree

A Git tree object creates the hierarchy between files in a Git repository. It lists the content of a git tree object
git ls-tree master

tree .git
It lists the content of the .git directory

6.28 git help

Display helpful information about Git

git --help

Chapter 7
Branching and Merging Strategy

The existing version control systems (VCS) entails to same kind of branching and merging strategy. We must know the term of branching which means that you create a separate new line of code and deploy to separate container for your work and continue to work there.

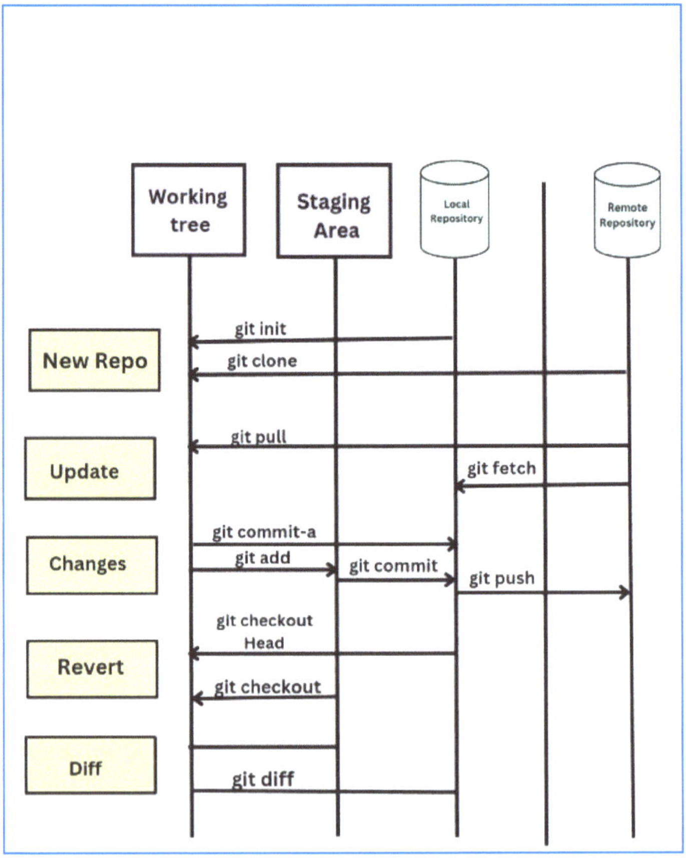

Figure: A holistic view of Git working

Some of the important branching naming conventions are as:

master represents the current public release branch. It keeps the current code deployed at the production server.
The development branch is the common development branch for all developers and is generally created as develop
feature branches are grouped under feature/ and belong to a feature developed by a developer or group of developers.
release branch the is branch which s tagged as the release. It is merged from develop branch and ready for higher environment production deployment. The hotfix branch is the branch that is created to fix the production issue. The bug fix branch is created from a release branch and merged back to the release branch.

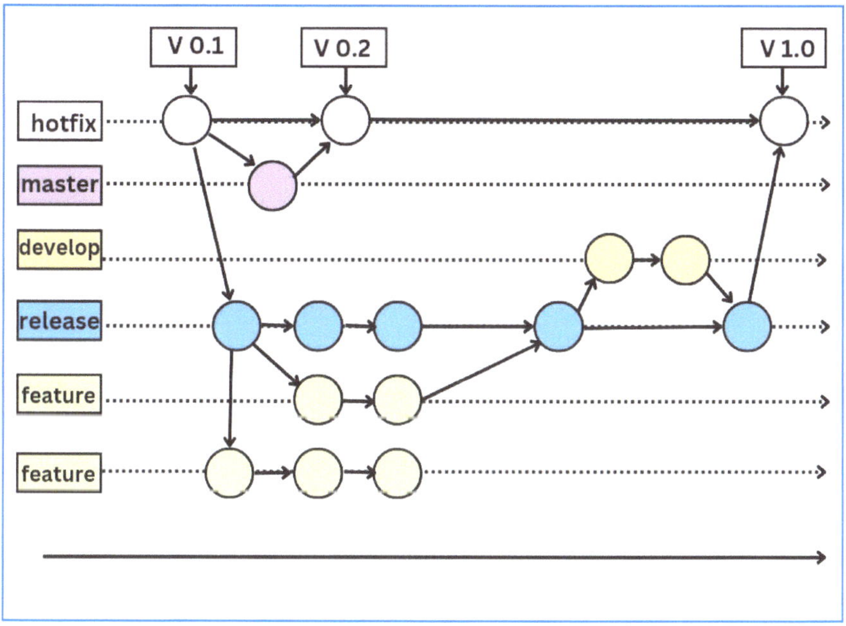

Figure: A typical Branching and Merging strategy in

A branching and merging strategy deviates from team to team and organization to organization.
But a good branching and merging strategy are that developer must create a feature branch always from develop branch. Once a feature branch or other features branches are ready then they should be

merged back to develop branch so that it has all the latest features to be deployed in the development environment.
Once we have deployed the code using CI/CD pipeline the development server and the code are tested by the testing team. Then testing team provides sign-off and requests to merge the develop branch code to the release branch.

At this time again CI/CD pipeline creates a release build and deploys to SIT/QA server.

Once code is deployed at QA/SIT server then QA testing performs several types of testing like regression testing, performance testing, or integration testing. If there is a bug found on the release then the developer creates a bugfix branch from the release branch and fixes the code in that bugfix branch. Once this build is fully signed off by testers, this bugfix branch is merged back to the release branch and develop branch.

This time, A release engineer creates the release tag at this build in VCS like git. Then this release is considered a candidate for the production server and deployed at the production server by CI/CD pipeline after a downtime approval from stakeholders. If there is a bug found on the production server then a hotfix branch is created from the master branch to fix the urgent production issue. This branch is built and deployed to the QA server. Once it is tested by developers at the QA server the code is again deployed to the production server.

Finally, the hotfix branch is merged back into the production branch, release branch and develop branch.

7.1 Conventions in Branching

A git branch should start with a category. Pick one of these: feature, bugfix, hotfix, or test.

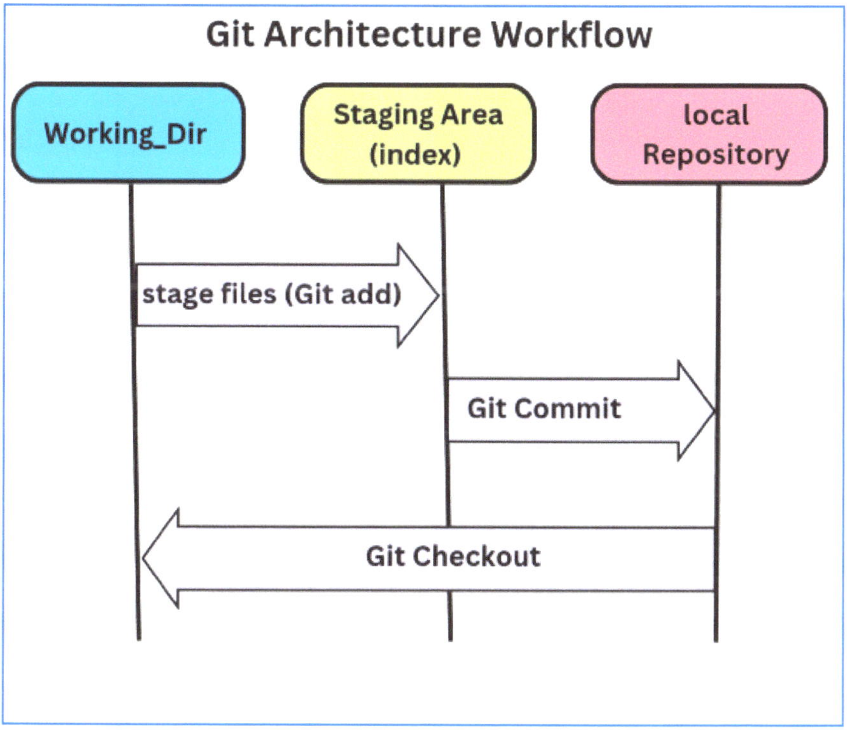

Figure: A flow of commit from Working directory to Remote via Staging

The feature is for adding, refactoring, or removing a feature
The bugfix is for fixing a bug

A hotfix is for changing code with a temporary solution and/or without following the usual process (usually because of an emergency) a kind of conventions

- The test is for experimenting outside of an issue/ticket
- Add a short descriptor of the task
- Start the branch name with a Group word
- Use Unique ID in branch names
- Use issue tracker IDs in branch names
- Use Hyphen or Slash as Separators
- Do not use bare numbers as leading parts.
- Git Branch with Author Name
- Avoid long descriptive names for long-lived branches

7.2 What is Merge Conflict

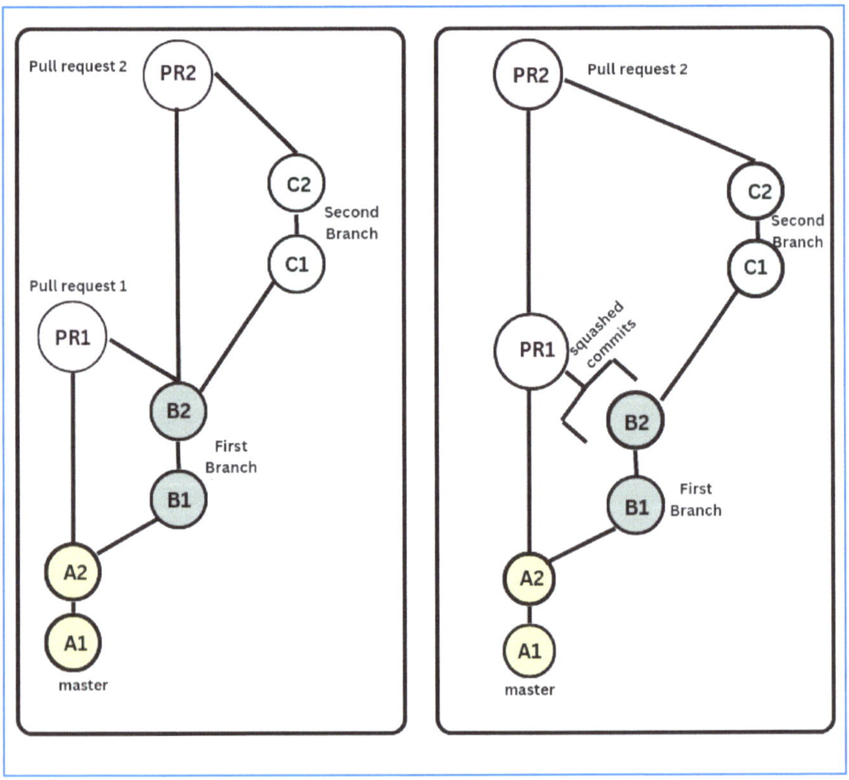

Figure: A view of Merge conflict situation

Merge conflict is a critical state which happens when the same lines in a file are being touched by two or more developers, or if one developer deleted a file while another developer was modifying it. In these cases, Git cannot automatically determine what line of code is wrong and which line of code needed to be considered. It means that Git is unable to automatically resolve differences in code between two commits.

```
$ git merge BRANCH-NAME
> Auto-merging styleguide.md
> CONFLICT (content): Merge conflict in styleguide.md
> Automatic merge failed; fix conflicts and then commit the result
```

Figure: Git Merge command

A merge can enter a conflicted state in two separate situations.

First Situation: Git fails to start the merge

This merge conflict happens because of conflicts with pending local changes by the same developer at his local and not because of conflicts with other developers,

This situation needs to be addressed using git stash, git checkout, git commit, or git reset.

Error Message:
You can see the error like this *Entry '<filename>' not UpToDate. Cannot merge. (Changes in working directory)*

Second Situation:Git fails during the merge

A failure of merge happens between the current local branch and the branch being merged.

Error Message:
error: Entry '<filename>' would be overwritten by merge. Cannot merge. (Changes in staging area)

7.3 How to Resolve Merge Conflict

Merge conflict must be resolved to support continuity of development efforts and final product. Here a merge conflict is to edit the conflicted file and then retry to merge the code. Developer can modify the file and then execute the git add command to stage the new merged content.

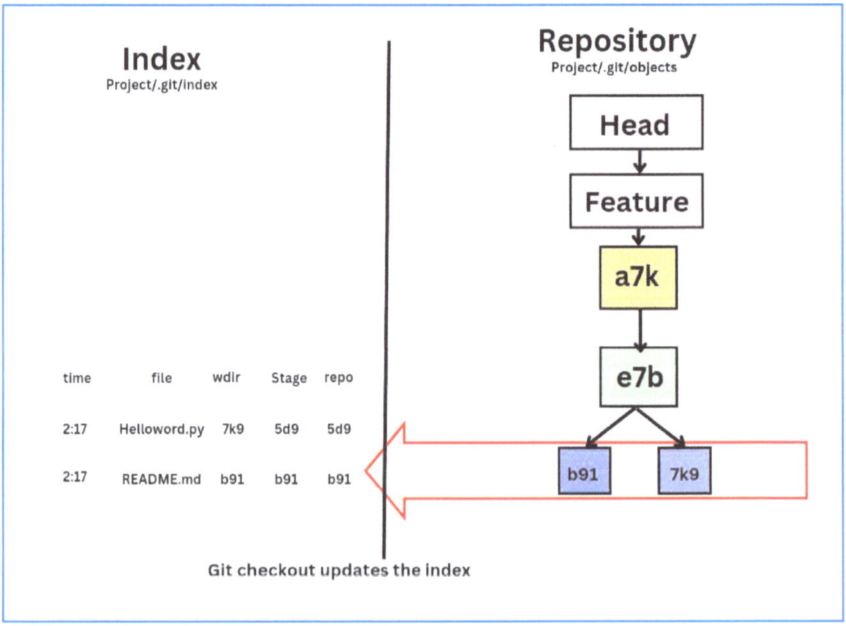

Figure: Indexing and repository

Use Git Commands to resolve Merge Conflict

git status
The status command is used when working with Git and during a merge, it will help identify conflicted files.

git log --merge
If we use the –merge option with the git log command, it will give us a list of commits that conflict between the merging branches.

git diff
This command will help to find differences between states of a repository/files.

git reset --mixed
The reset command can be used to undo changes to the working directory and staging area.

git merge --abort
Executing git merge with the --abort option will exit from the merge process and return the branch to the state before the merge began.

git stash
git stash command is to store the unfinished local work temporarily. Git-stash gives you a clean repo and reverts the repo state to the last commit. Then we can use git pull to pull the updates.

Chapter 8
Security in GIT

Figure: Security requirement is essential

Several organizations use different kinds of GitLab, GitHub, or a self-hosted Git server. It has been seen that several vulnerabilities like CVE-2022-24765 or CVE-2022-24767 sneak up and start a negative effect of unpleasant repercussions.
At its core, Git is designed more for collaboration and not built for security.

Please refer to some vulnerabilities found in Git recently https://github.blog/2022-04-12-git-security-vulnerability-announced .

Reasons for vulnerabilities in Version Control System

- Developers commit sensitive information
- Unsigned commits and Inaccurate access permissions
- Insecure Directories (.git/config)
- Multi-Factor Authentication
- Git and related tools are outdated
- Insecure pipeline configuration
- Unpatched software

There are several methods to avoid some common pitfalls when it comes to Git security.

Developers should not commit sensitive information

Keeping sensitive information in Git can lead to vulnerable condition and happens more often than you might think. Private keys should be ignored by Git (*.pem, *.key,) but other sensitive files are harder to anticipate. Git use GPK for securing sensitive files.

Unsigned commits and Inaccurate access permissions

Git enables you to cryptographically sign your work using GPG keys. It provides you with different access levels to log in to your project. If Git admin does not provide the right access to developers or weak access then it leads to making Git less secure and a candidate to get vulnerabilities. Sometimes inaccurate access permissions cause to Git be insecure. Git admins have to manage the repository settings and do the correct setting for repository visibility.

Insecure Directories (.git/config)

Internet cyber criminals and hackers send some URLs containing the git directory to access the metadata within a Git repository. Metadata will have several useful information like customer data information which leads to hacking.

Figure: Index of .git folder

Multi-Factor Authentication

Many security breaches can be avoided by strengthening security measures, such as using two-factor authentication. Multi-Factor authentication like pinged can help to protect Git.

Git and related tools are outdated

The recent CVE-2018-11235 is a vulnerability for Git that has drawn attention worldwide. Updating Git libraries from time to time helps to secure Git.

Insecure pipeline configuration

An insecure CI/CD pipeline can lead to secrets leaking. For example, as pull requests coming from forks of developers' repositories or CI/CD processes left operating unattended. Linting is a process performed by tools that identify coding style deviations and unsafe practices. synthetic data, service virtualization, and data masking Triggering penetration tests to find back doors and other entry-point vulnerabilities.

Unpatched software

It is always recommended to have Git software to be updated from time to time with sophisticated libraries.

Git and Cryptography

In cryptography, SHA-1 (Secure Hash Algorithm 1) is a cryptographic hash function that takes an input and produces a 160-bit (20-byte) hash value known as a message digest – typically rendered as a hexadecimal number, 40 digits long.

Chapter 9
GIT TOOLS

Git comes with some of its tools like Git Bash, and Git GUI which provides a way to better interact with remote repositories and handle several complex situations. It supports inbuilt as well as third-party tools. Git supports various important tools like git bash, git-gui, and gitk for committing and browsing.

9.1 GitBash

Git Bash is an application for the Windows environment. It is used as a Git command line for windows.

9.2 Git GUI

Git GUI is a powerful alternative to Git BASH. It offers a graphical feature of the Git command line functions. It is a tool that allows the a GUI to interact with visual mode. It advocate for GUI interaction rather than writing commands manually, offering a convenient graphical interface with in-built options.

9.3 Gitk

Gitk is a graphical history viewer tool. Gitk is a graphical repository browser. GitK can be considered as a wrapper layer on the top of Git for exploring the git logs.Either reload gitk from the command line or use the GUI and navigate to File -> Reload.

gitk [<options>] [<revision range>] [--] [<path>...]

Here are some most common options with gitk.
--all
Show all refs (branches, tags, etc.).
--remotes[=<pattern>]
--branches[=<pattern>]
--tags[=<pattern>]

Assume that all the branches (tags, remote branches, resp.) are listed on the command line as <commit>. If <pattern> is given, limit refs to ones matching given the shell glob. If the pattern lacks ?, *, or [, /* at the end is implied.

Some popular third-party tools

9.4 TortoiseGit

TortoiseGit is a free software Git revision control client, implemented as a Windows shell extension and based on TortoiseSVN. TortoiseGit significantly facilitates and simplifies your interaction with Git. It operates as a standalone application and as a Microsoft Windows shell extension.

9.5 GitHub Desktop

GitHub Desktop is an open-source tool that enables you to be more productive. GitHub Desktop encourages you and your team to collaborate using best practices with Git and GitHub.

9.6 Sublime Merge

Sublime Merge is a Graphical Git Client from the creators of Sublime Text. The syntax highlighting engine is used in Sublime Text.

9.7 Eclipse Git Plug-In

Eclipse is a popular cross-browser IDE written in Java. It has a great integration with Git.
For this kind of integration, a Git plug-is required to be placed inside the developer's IDE.
It provides basic features for developers on day to day operations like commit, amend commits, view diffs, and
visualize the repository history graphically.

9.8 gitweb

gitweb is a web interface for any operations to Git repositories. It is written in Perl and can be used as a CGI script. Gitweb helps in browsing a git repository using a web browser.

9.9 VCSCommand

VCSCommand is a plug-in used for Vimeditor. It's written in Python. Through VCSCommand you can execute several operations with several VCS directly through Vim.

Chapter 10
Advanced Topics

10.1 What are Git Hooks

Git hooks are shell scripts found in the hidden dot git/hooks directory of a Git repository. These scripts are used to accomplish some actions in response to specific events after triggered . It is a kind of automation of your development efforts and reduce manual intervention in a development lifecycle.

Hooks are implemented to address the local policies and execute some actions on some events. It means they are purely event-based. The hooks reside in a directory within the VCS repository to see if there is an associated script to run.

Hooks are generally implemented in scripts that improve functionality and enhance the restriction policies based on different events and actions. Within .git/hooks/, there are several sample hooks available such as post-update.sample,pre-commit, and pre-receive.sample, which are good examples.

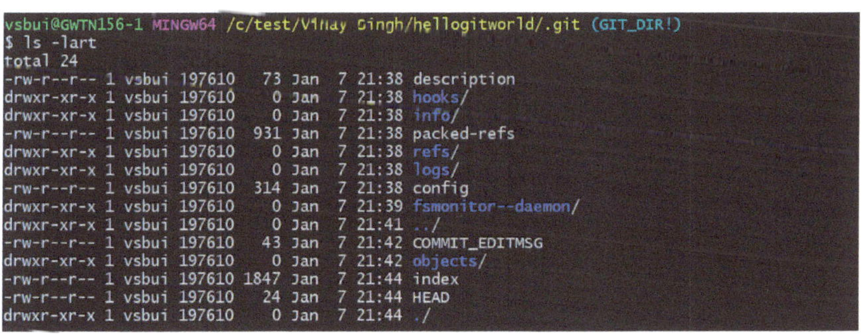

```
vsbui@GWTN156-1 MINGW64 /c/test/Vinay Singh/hellogitworld/.git (GIT_DIR!)
$ ls -lart
total 24
-rw-r--r-- 1 vsbui 197610   73 Jan  7 21:38 description
drwxr-xr-x 1 vsbui 197610    0 Jan  7 21:38 hooks/
drwxr-xr-x 1 vsbui 197610    0 Jan  7 21:38 info/
-rw-r--r-- 1 vsbui 197610  931 Jan  7 21:38 packed-refs
drwxr-xr-x 1 vsbui 197610    0 Jan  7 21:38 refs/
drwxr-xr-x 1 vsbui 197610    0 Jan  7 21:38 logs/
-rw-r--r-- 1 vsbui 197610  314 Jan  7 21:38 config
drwxr-xr-x 1 vsbui 197610    0 Jan  7 21:39 fsmonitor--daemon/
drwxr-xr-x 1 vsbui 197610    0 Jan  7 21:41 ../
-rw-r--r-- 1 vsbui 197610   43 Jan  7 21:42 COMMIT_EDITMSG
drwxr-xr-x 1 vsbui 197610    0 Jan  7 21:42 objects/
-rw-r--r-- 1 vsbui 197610 1847 Jan  7 21:44 index
-rw-r--r-- 1 vsbui 197610   24 Jan  7 21:44 HEAD
drwxr-xr-x 1 vsbui 197610    0 Jan  7 21:44 ./
```

Figure: A view of repository with hooks folder

```
vsbui@GWTN156-1 MINGW64 /c/test/Vinay Singh/hellogitworld/.git/hooks (GIT_DIR!)
$ ls -lart
total 53
-rwxr-xr-x 1 vsbui 197610  478 Jan  7 21:38 applypatch-msg.sample*
-rwxr-xr-x 1 vsbui 197610  896 Jan  7 21:38 commit-msg.sample*
-rwxr-xr-x 1 vsbui 197610 4655 Jan  7 21:38 fsmonitor-watchman.sample*
-rwxr-xr-x 1 vsbui 197610  189 Jan  7 21:38 post-update.sample*
-rwxr-xr-x 1 vsbui 197610  424 Jan  7 21:38 pre-applypatch.sample*
-rwxr-xr-x 1 vsbui 197610 1643 Jan  7 21:38 pre-commit.sample*
-rwxr-xr-x 1 vsbui 197610  416 Jan  7 21:38 pre-merge-commit.sample*
-rwxr-xr-x 1 vsbui 197610 1374 Jan  7 21:38 pre-push.sample*
-rwxr-xr-x 1 vsbui 197610 4898 Jan  7 21:38 pre-rebase.sample*
-rwxr-xr-x 1 vsbui 197610  544 Jan  7 21:38 pre-receive.sample*
-rwxr-xr-x 1 vsbui 197610 1492 Jan  7 21:38 prepare-commit-msg.sample*
-rwxr-xr-x 1 vsbui 197610 2783 Jan  7 21:38 push-to-checkout.sample*
drwxr-xr-x 1 vsbui 197610    0 Jan  7 21:38 ./
-rwxr-xr-x 1 vsbui 197610 3650 Jan  7 21:38 update.sample*
drwxr-xr-x 1 vsbui 197610    0 Jan  7 21:44 ../
```

Figure: List of files in hooks

Generally, Hooks are of two types:

- Local Hooks
- Server-Side Remote Hooks

Some examples of hooks are:

- pre-commit: It occurs before a git commit is accepted.
- post-commit: Occurs after a git commit is accepted.
- post-checkout: It occurs before a git checkout.
- pre-rebase: It occurs before a git-rebase.

pre-commit hook:

The pre-commit script is executed every time you run git commit before Git looks for developer for a commit message or create a commit object. It gets executed before a git commit command is taken place. For example, a restriction or policy can be applied for developers to make sure they mandatory mention their Jira story number in a commit message.

post-commit hook:

This hook is invoked by git-commit. It does not take any parameters. It is invoked after a commit is made and usually calls a CI/CD pipeline to support build automation.

pre-rebase

This hook is called git-rebase. It is generally used to stop rebase in early stage before actual rebase is executed. It means it prevent a branch from getting rebased. The hook allows you to add two parameters. The upstream is called the first parameter from which the series was forked. The branch is the second parameter which happens while being rebased and is not set when rebasing the current branch.

post-checkout

On the situation of git-checkout or git-switch is run after having updated the work tree, then post-checkout hook is invoked. The hook is given three parameters:

- the ref of the previous HEAD,
- the ref of the new HEAD
- a flag indicating whether the checkout was a branch checkout or a file checkout.

It is also run after git-clone, unless the --no-checkout (-n) option is used. It is leveraged to accomplish the auto-display differences from the previous HEAD if different, set working dir metadata properties or validity checks for the repositories.

post-merge

git-merge invokes this hook on a very special situation when git pull is done on a local repository. A status flag is a single parameter. This hook cannot affect the outcome of git merge and is not executed if the merge failed due to conflicts.

pre-push

git-push calls this kind of pre-push and can be used to prevent a push from taking place. The hook is called with two parameters which provide the name and location of the destination remote.

Steps to Implement the pre-commit hook

- Clone the Repository
- Create a Directory

At the repository root level, Create a directory for holding the scripts (let's say that the name will be: git-hooks)

```
user@user-PC MINGW64 ~/git-test/HelloWorld (master)
$ cd git-hooks/

user@user-PC MINGW64 ~/git-test/HelloWorld/git-hooks (master)
$ ls -lart
total 9
-rwxr-xr-x 1 user 197121 347 Sep 23 17:22 commit-msg*
drwxr-xr-x 1 user 197121   0 Sep 23 21:00 ./
drwxr-xr-x 1 user 197121   0 Sep 23 21:00 ../

user@user-PC MINGW64 ~/git-test/HelloWorld/git-hooks (master)
$
```

Figure: go to git-hooks

```
MINGW64:/c/Users/user/git-test/HelloWorld/git-hooks
#!/bin/bash

MSG_FILE=$1
FILE_CONTENT="$(cat $MSG_FILE)"

export REGEX='(TOOL-[0-9]{4,5}|Merge)( .+)?'
export ERROR_MSG="Error: commit message format must match regex \"${REGEX}\""

if [[ $FILE_CONTENT =~ $REGEX ]]; then
 echo "Good commit message."
else
 echo $ERROR_MSG
 echo "Actual commit message: \"$FILE_CONTENT\""
 exit 1
fi
~
```

Figure: Script written for git hooks

Add this file to the master so that when the developer creates a feature branch he should have this folder and file in his feature branch.

- Execute few commands
  git add .
  git commit -m "TOOL-0000 Adjusting the hook"
  git push origin master
  git checkout feature-1

- Also, thee commands
  cd ../
  git config core.hooksPath=./git-hooks
  git config core.hooksPath ./git-hooks

Result:
This will help to apply the policy that no developer will allow to commit a message without TOOL-**** phrase in a commit message.This command instructs (only for this single repository) git that the scripts acting as hooks are stored in the git-hooks directory.

From now on if you try to commit anything with the commit message not starting with TOOL-XXXX or TOOL-XXXXX (where X is a digit) the commit will fail with the proper message.

10.2 What are Objects and BLOBS

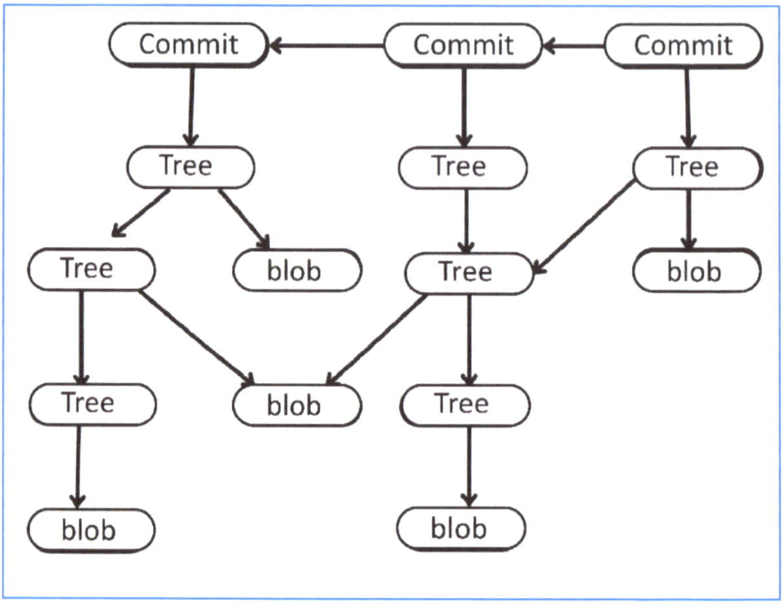

Figure: A view of BLOB and TREE

Git stores different types of objects in .git/objects.

The object types are:

- commit
- Tree
- Blob
- tag

```
vsbui@GWTN156-1 MINGW64 /c/test/Vinay Singh/hellogitworld/.git (GIT_DIR!)
$ ls -lrt
total 16
-rw-r--r-- 1 vsbui 197610   73 Jan  7 21:38 description
drwxr-xr-x 1 vsbui 197610    0 Jan  7 21:38 hooks/
drwxr-xr-x 1 vsbui 197610    0 Jan  7 21:38 info/
-rw-r--r-- 1 vsbui 197610  931 Jan  7 21:38 packed-refs
drwxr-xr-x 1 vsbui 197610    0 Jan  7 21:38 refs/
drwxr-xr-x 1 vsbui 197610    0 Jan  7 21:38 logs/
-rw-r--r-- 1 vsbui 197610  314 Jan  7 21:38 config
drwxr-xr-x 1 vsbui 197610    0 Jan  7 21:39 fsmonitor--daemon/
-rw-r--r-- 1 vsbui 197610   43 Jan  7 21:42 COMMIT_EDITMSG
drwxr-xr-x 1 vsbui 197610    0 Jan  7 21:42 objects/
-rw-r--r-- 1 vsbui 197610 1847 Jan  7 21:44 index
-rw-r--r-- 1 vsbui 197610   24 Jan  7 21:44 HEAD

vsbui@GWTN156-1 MINGW64 /c/test/Vinay Singh/hellogitworld/.git/objects/e8 (GIT_DIR!)
$ git log
commit e837eca26c45464a6d01849e8004b47930151e3d (HEAD -> develop, master)
```

Figure: A list of your repository

git hash-object stores data in .git/objects directory and gives you back the unique key.

```
vsbui@GWTN156-1 MINGW64 /c/test/Vinay Singh/hellogitworld/.git/objects (GIT_DIR!
)
$ ls -ltt
total 4
drwxr-xr-x 1 vsbui 197610 0 Jan  7 21:42 e8/
drwxr-xr-x 1 vsbui 197610 0 Jan  7 21:42 a1/
drwxr-xr-x 1 vsbui 197610 0 Jan  7 21:38 pack/
drwxr-xr-x 1 vsbui 197610 0 Jan  7 21:38 info/
```

Figure: A list of directories of objects inside object folder

Commit object type

The commit object contains the directory tree object hash, parent commit hash, author, committer, date, and message.

```
vsbui@GWTN156-1 MINGW64 /c/test/Vinay Singh/hellogitworld/.git/objects/a1 (GIT_DIR!)
$ ls -lrt
total 1
-r--r--r-- 1 vsbui 197610 366 Jan  7 21:42 ea4079d8317de74e28bf596c3ad13397bef9a8

vsbui@GWTN156-1 MINGW64 /c/test/Vinay Singh/hellogitworld/.git/objects/a1 (GIT_DIR!)
$ cd ../e8/

vsbui@GWTN156-1 MINGW64 /c/test/Vinay Singh/hellogitworld/.git/objects/e8 (GIT_DIR!)
$ ls -lrt
total 1
-r--r--r-- 1 vsbui 197610 187 Jan  7 21:42 37eca26c45464a6d01849e8004b47930151e3d
```

Figure: A view of subfolder inside object folder

git cat-file -t shows us the type of the object represented by a particular hash

```
vsbui@GWTN156-1 MINGW64 /c/test/Vinay Singh/hellogitworld/.git/objects/e8 (GIT_DIR!)
$  git cat-file -t e837eca26c45464a6d01849e8004b47930151e3d
commit
```

Figure: git cat-file command with commit

git cat-file -p shows the contents of the file associated with this hash

```
vsbui@GWTN156-1 MINGW64 /c/test/Vinay Singh/hellogitworld/.git/objects/e8 (GIT_DIR!)
$ git cat-file -p e837eca26c45464a6d01849e8004b47930151e3d
tree a1ea4079d8317de74e28bf596c3ad13397bef9a8
parent ef7bebf8bdb1919d947afe46ab4b2fb4278039b3
author sendmequickfast@gmail.com <sendmequickfast@gmail.com> 1673152948 -0700
committer sendmequickfast@gmail.com <sendmequickfast@gmail.com> 1673152948 -0700

Adding a python script to do VCS Migration
```

Figure: A result of git cat-file command

Tree object type

A Git tree object creates the hierarchy between files in a Git repository. You can use the Git tree object to create the relationship between directories and the files they contain.

The commit contents give us the hash of the directory listing for the commit. If we inspect this object, we find it is of type “tree” and contains the directory listing for the commit.

```
vsbui@GWTN156-1 MINGW64 /c/test/Vinay Singh/hellogitworld/.git/objects/e8 (GIT_DIR!)
$ git cat-file -t  a1ea4079d8317de74e28bf596c3ad13397bef9a8
tree
```

Figure: A tree object

Blob object type

A Git blob is called binary large object. It is the object type used to store the contents of each file in a repository. The file's SHA-1 hash is computed and stored in the blob object.

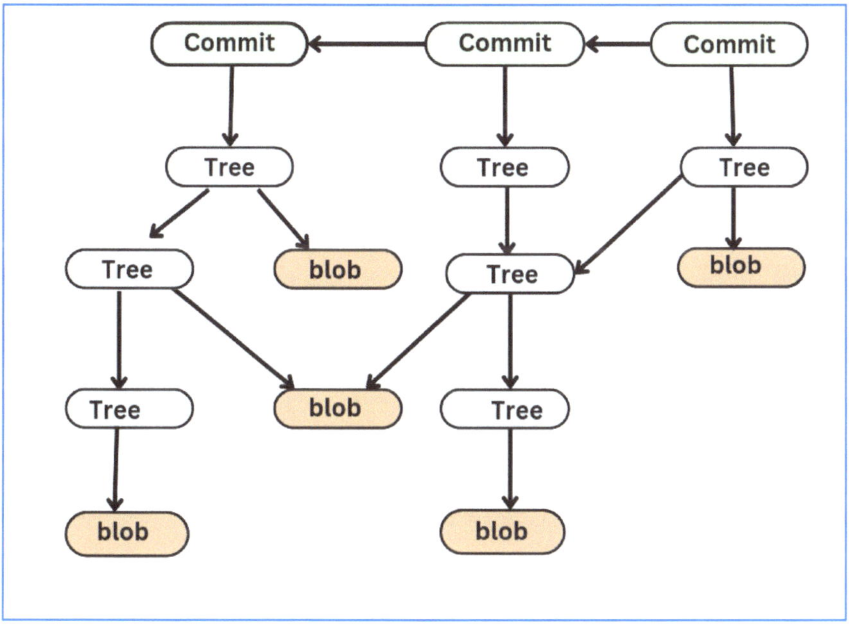

Figure: A view of BLOB and TREE objects

The directory listing gave us the hash of the stored in the file.

This object is of type "blob" and contains the file snapshot.

Figure: A view of BLOB object

Tag object type

A Git tag is tagging of a commit to be used in specific release deployment .It is similar to a Git reference, but the Git commit that it points to never changes. Git tags are used in the situation where a specific release is ready for production and need a build reference.

There is another object type which is a tag object.

```
vsbui@GWTN156-1 MINGW64 /c/test/Vinay Singh/hellogitworld/.git/objects/e8 (GIT_DIR!)
$ git tag -a first-commit -m "my first tag"
```

Figure: A git tag command

It will create another new object in .git/objects.

```
vsbui@GWTN156-1 MINGW64 /c/test/Vinay Singh/hellogitworld/.git/objects (GIT_DIR!)
$ ls -lrt
total 4
drwxr-xr-x 1 vsbui 197610 0 Jan  7 21:38 info/
drwxr-xr-x 1 vsbui 197610 0 Jan  7 21:38 pack/
drwxr-xr-x 1 vsbui 197610 0 Jan  7 21:42 a1/
drwxr-xr-x 1 vsbui 197610 0 Jan  7 21:42 e8/
drwxr-xr-x 1 vsbui 197610 0 Jan 31 15:17 03/

vsbui@GWTN156-1 MINGW64 /c/test/Vinay Singh/hellogitworld/.git/objects (GIT_DIR!)
$ cd 03

vsbui@GWTN156-1 MINGW64 /c/test/Vinay Singh/hellogitworld/.git/objects/03 (GIT_DIR!)
$ ls -lrt
total 1
-r--r--r-- 1 vsbui 197610 141 Jan 31 15:17 46c40e2d1c9fae39d260877a093f4f806331cb
```

Figure: A view of your object directory

10.3 What are Hashes

At a high level, a git commit hash is a SHA1 hash of the state of the git repository at the time of the commit. Every commit to a repository has a unique identifier called a hash. Hashes facilitate git by sharing the data efficiently between several repositories. Here is a point if two files are the same, their hashes are definitely to be the same. Similarly, if two commits contain the same files and have the same parent, their hashes will be the same as well.

10.4 What is Indexing

The Git index is a critical data structure in Git. Indexing helps to get the as the "staging area" between the files you have on your filesystem and your commit history.

The index is a binary file, and it is kept in a special folder called .git/index. This folder contains a sorted list of path names, each with permissions and the SHA1 of a blob object. Git never saves its objects into a directory or folder. Instead, it stores information about each file in our repository.

mtime — is the time of the last update
file — name of the file
wdir —file version in the working directory
stage —file version in the index
repo — file version in the repository

Git Index may be defined as the staging area between the workspace and the repository. Git index helps to set up all changes before you commit them to your local repository and finally combine them.

Command:

git ls-files --stage can show you the contents of the index

```
vsbui@GWTN156-1 MINGW64 /c/test/Vinay Singh/hellogitworld/.git/objects/03 (GIT_DIR!)
$ git ls-files
.gitattributes
.gitignore
.travis.yml
README.txt
SVNtoGIT.py
build.gradle
fix.txt
pom.xml
resources/labels.properties
runme.sh
src/Division.groovy
src/Main.groovy
src/Square.groovy
src/Subtract.groovy
src/Sum.groovy
src/main/java/com/github/App.java
src/test/java/com/github/AppTest.java

vsbui@GWTN156-1 MINGW64 /c/test/Vinay Singh/hellogitworld/.git/objects/03 (GIT_DIR!)
$ git ls-files --stage
100644 412eeda78dc9de1186c2e0e1526764af82ab3431 0       .gitattributes
100644 7c820833a9ad5fbfc96efd533d55f5edc65dc977 0       .gitignore
100644 17eccd68b7cafa718d53c8b4db666194646e2bd9 0       .travis.yml
100644 adfd6da3c0a3fb038393144becbf37f14f780087 0       README.txt
100644 e69de29bb2d1d6434b8b29ae775ad8c2e48c5391 0       SVNtoGIT.py
100644 6058be211566308428ca6dcab3f08cf270cd9568 0       build.gradle
100644 e69de29bb2d1d6434b8b29ae775ad8c2e48c5391 0       fix.txt
100644 99975710477a65b89233b2d12bf60f7c0ffc1f5c 0       pom.xml
100644 fbdcfef007c0c09061199e687087b18c3cf8e083 0       resources/labels.properties
100755 fb4bd4ec9220ed4fe0d9526d1b77147490ce8842 0       runme.sh
100644 3862298eb2efbc26cd8def50e72c66b1e03097db 0       src/Division.groovy
100644 6191a102912c9786fdfcb1ce052afcaf386d6e36 0       src/Main.groovy
100644 fde331951be7ee45857487ffff1ff50ad8c90d7d 0       src/Square.groovy
100644 726e5252ccafd9d8fc661ed5ca5c665f24d4bc22 0       src/Subtract.groovy
100644 9f221918607d78c1db3f6c6d5afa68a66b1146a8 0       src/Sum.groovy
100644 6f77f90fba04ad0be7e34a8c4fca20a65a1e518a 0       src/main/java/com/github/App.java
100644 43767197a768385d97ce751c421ee9e7ceeda5a7 0       src/test/java/com/github/AppTest.java
```

Figure: A command git ls files

10.5 Handling Empty Repositories

The problem with Git is that it does not entertain empty repositories. It means a developer cannot push the empty repository into remote Git.

In order to get Git to recognize an empty directory, the unwritten rule is to put a file named .gitkeep in it. Git will see the .gitkeep file in the otherwise empty folder and make that folder part of the next commit or push.

Follow these steps to add a placeholder file like .gitkeep to remote Git:

- Create an empty folder with the mkdir command
- Create the .gitkeep file with the touch .gitkeep command

- Use git add . to update the Git index
- Perform a git commit lime git commit -m "adding .gitkeep file"
- Push the commit with the empty Git directory to remote repositories

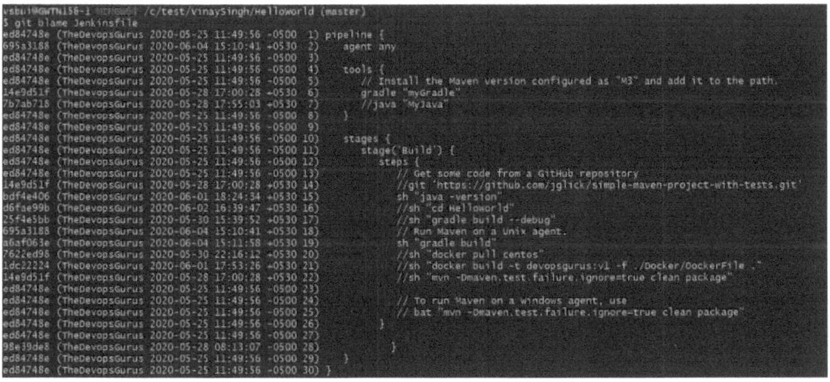

Figure: A command git ls files

10.6 What is Git Blame

The git blame command is a versatile troubleshooting utility that has extensive usage options like it display of author metadata attached to specific committed lines in a file.

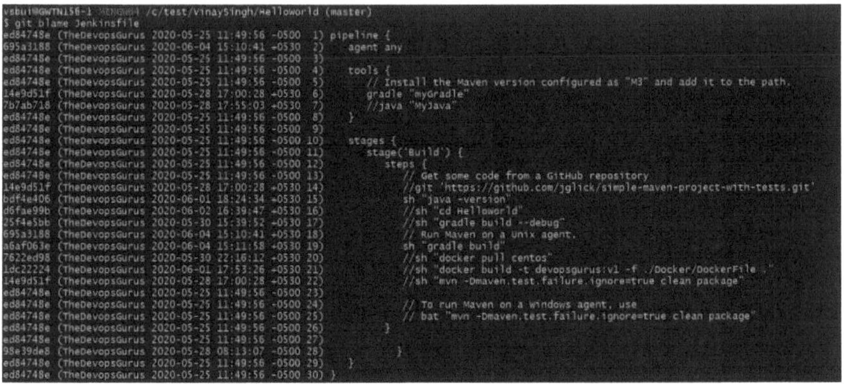

Figure: A view of git blame command

The git blame displays the last author who changed a line. Sometimes it is cumbersome with git blame to know when a line was initially added. Hence Git Blame comes with some options like *-w, -c, and -m* options can be used. But it will be easier to use the git log.

git blame -w – Simplifies output, so whitespace changes made by an author aren't listed in the revisions
git blame -c – Identifies lines that are copied from other files and displays the original author metadata
git blame -e – Instead of displaying the author's username, the email address is displayed
git blame -m – Identifies lines that have been moved or copied from within the same file. The original author is reported, not the author that relocated that line.
git blame -l – Will display commit hashes at full-length
git blame -t – Displays time and date as a timestamp instead of the default date and time format.
Developers preferably use git log with the -S option to display all initial commits with a particular added or changed code.

Cherry-picking in Git means choosing a commit from one branch and applying it to another.

As with other Git operations that introduce changes via the process of applying a diff, you may need to resolve conflicts to fully apply the changes from the given commit. The git cherry-pick command is one you use to incorporate changes from other branches into your current branch.

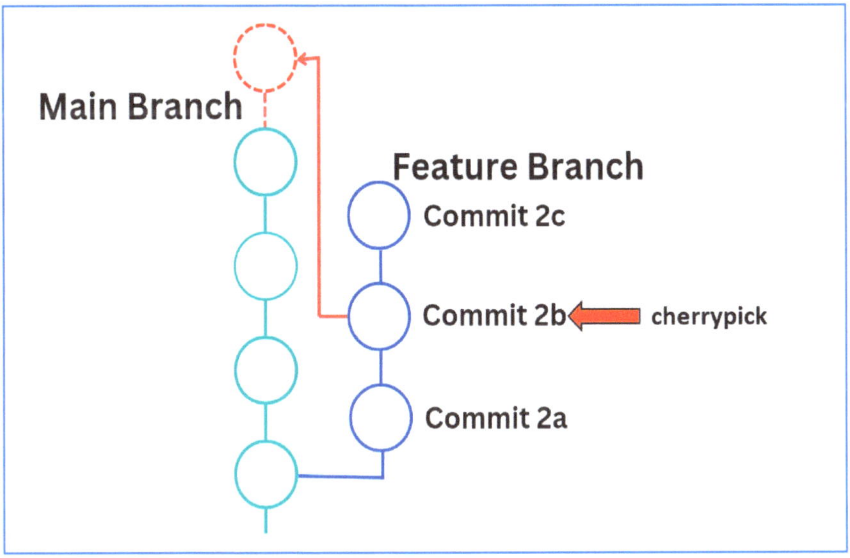

Figure: A cherry pick command

So, Cherry picking is an easy way to grab a commit from a branch and apply it to another. git cherry-pick can be useful for undoing changes. When merging, you'll typically want to incorporate all the changes made in the other branch into the current one.

But sometimes you want only some of the changes. In these cases, cherry-picking comes in handy

For example, say a commit is accidentally made to the wrong branch. You can switch to the correct branch and cherry-pick the commit to where it should belong. Sometimes it is very dangerous if used without mindfulness as it can lead to duplicate commits. Cherry picking should not be misused in place of git merge or git rebase. Sometimes developers make some commits from 2 or more branches update similar lines of code with various substances and git cherry-pick one commit to the other branch, which prompts conflict as well.

The use case of cherry-picking is :

- Undoing changes and restoring lost commits
- Team collaboration
- Bug hotfixes

Execute these commands to see how a cherry-pick command works:
git log --all --decorate --oneline –graph
git cherry-pick revision number

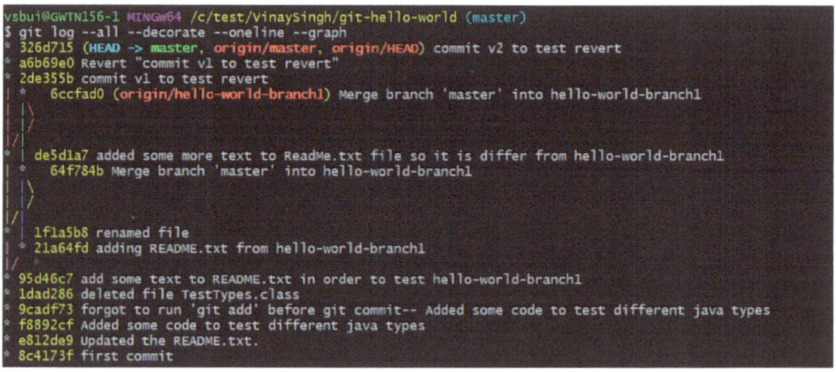

Figure: Git log to show how a cherry-pick command works

```
vsbui@GWTN156-1 MINGW64 /c/test/VinaySingh/git-hello-world (master)
$ git cherry-pick 326d715
The previous cherry-pick is now empty, possibly due to conflict resolution.
If you wish to commit it anyway, use:

    git commit --allow-empty

Otherwise, please use 'git cherry-pick --skip'
On branch master
Your branch is up to date with 'origin/master'.

You are currently cherry-picking commit 326d715.
  (all conflicts fixed: run "git cherry-pick --continue")
  (use "git cherry-pick --skip" to skip this patch)
  (use "git cherry-pick --abort" to cancel the cherry-pick operation)

nothing to commit, working tree clean
```

Figure: A result of cherry-pick command

10.7 Reset, Revert, and Rebase

There are several situations multiple times where developers need to roll back to one of their previous commits.

Git reset

Git reset is a complex command, and it is used to undo the changes like a rollback. When you use the Git reset feature, it will return your current environment to the previous commit.

There are three modes of running a git reset command: –soft, –mixed, and –hard. The reset command moves the branch pointer back in the chain (typically) to "undo" changes

The default mode is the mixed mode.

Three options are shown below:

git reset --hard

It allows moving the HEAD to the specified commit. It will remove all the commits with happened after the specified commit.

```
$ git reset --hard HEAD
```

Figure: git reset command

git reset --soft

Git reset –soft is used to put HEAD back to a specified commit. It actually reset all the changes between current HEAD position and the specified commit and saves all the changes in the index.

In other words, Git reset --soft the changes as staged, ready to be committed again.

```
$ git reset --soft HEAD
```

Figure: A git reset command with soft option

git reset --mixed

It is almost same as Git reset soft. It used to perform a Git reset with the --mixed option will undo all the changes between HEAD and the specified commit. The only things is that it saves your changes in the Working Directory

git reset HEAD~3

git revert

It is used to perform a revert operation, i.e., to revert some changes. It allows a new commit to reverting to a particular commit. The revert command adds a new commit at the end of the chain to "cancel" changes.

The git revert command does not delete any data while performing the revert operation.

```
$ git revert HEAD
```

Figure: A git revert command with HEAD option

```
$ git revert 79b84c6
```

Figure: A git revert command with commit option

git rebase

It allows one to move or join the different commits of one branch over another branch. it is very similar to a merge.

git rebase feature

```
$ git rebase master
```

Figure: A git rebase command

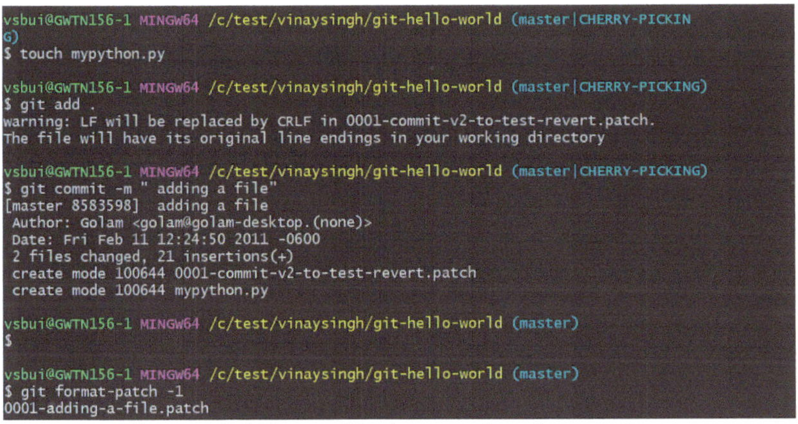

Figure: Adding a file in your local repository

10.8 What is GIT Patch

The patch is a text file. The contents of the patch are similar to Git diff, but along with code, it also has metadata about commits. Some of the metadata are like commit ID, date, commit message, etc.

To create a patch, we will use the git format-patch or git am bugfix Patch command.

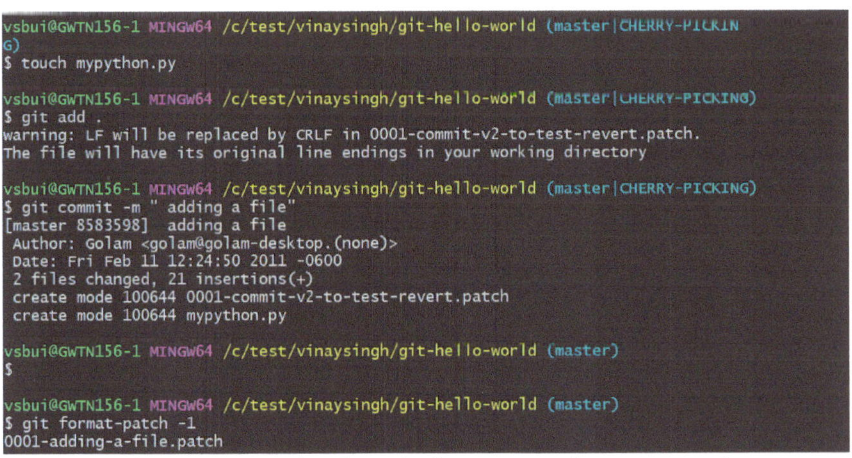

Figure: A result after git format-patch command

The above command creates .patch files inside the current working directory. Now the developer can use this patch to modify his files. There are two commands very useful .First is to apply patches git am and git format-patch. Git apply modifies the local files without creating a commit, while git am modifies the file and creates a commit as well.
Patches are an alternative way to exchange code changes.

10.9 What is SubModule

A Git submodule is used to integrate the several repositories into a single project. This is awesome feature and requires a subdirectory within the main project directory, but the code it contains does not have to be copied directly into the project.
Git submodules allow you to keep a git repository as a subdirectory of another git repository. Git submodules are kind of pointer to a repository which exhibits a snapshot in time. A git submodule is a record within a host git repository that points to a specific commit in another external repository.

Git submodules are a great feature as in agile culture where developers writes microservices and update each microservice separately to sustain independent change histories. Several projects can share the code maintained in a single repository by using Sub Module.

There are a few submodule commands:

```
$ git submodule add https://github.com/marko-pnap/new-submodule.git submodules/example
```

Figure: Git sub module command

```
vsbui@GWTN156-1 MINGW64 /c/test/vinaysingh/git-hello-world (master)
$ git submodule add https://github.com/marko-pnap/new-submodule.git submodules/example
Cloning into 'C:/Test/VinaySingh/git-hello-world/submodules/example'...
remote: Enumerating objects: 6, done.
remote: Counting objects: 100% (6/6), done.
remote: Compressing objects: 100% (4/4), done.
remote: Total 6 (delta 0), reused 0 (delta 0), pack-reused 0
Receiving objects: 100% (6/6), done.
warning: LF will be replaced by CRLF in .gitmodules.
The file will have its original line endings in your working directory
```

Figure: A git submodule command

git submodule
The command add a submodule to your main repository

git submodule init
The command registers the paths to submodules within the project tree.

git submodule update
It is used to clone those submodules which are missing and it fetches any new remote commits and updates the directory tree.

Chapter 11
GIT Administration

A git administrator role the most crucial role in the SDLC life cycle. All the code and their repository responsibility are on Git Administrator.

take up the administrative side of Git

Firewall Protection
Git uses a read-only protocol that may be blocked by your firewall (generally port 9418). Git Admin has to check on that.

Data recovery and troubleshooting
One of the most important things with a git infrastructure, as with any, is a correct and tested backup strategy. A backup without testing is next to useless.

Use Third-party tools

There are several tools available in the market that can be integrated with git and gives the development team more options and features to play conveniently rather than on the command line.
Some of them are :

- Gitfox
- Sourcetree
- GitHub Desktop
- Sublime Merge
- GitKraken Client
- SmartGit
- TortoiseGit

Repository Maintenance

Git administrator has to do some cleanup of a repository, usually to make it more compact. Such clean-ups are done by using *git gc --auto* command in each repository. This command checks whether there are too many loose objects.

Security Patches and Libraries

Git Administrator is responsible to apply security patches against some vulnerabilities and keep them up-to-date.

User Access

Git administrator provides access to different users on different projects based upon their role.

Chapter 12
SVN to GIT Migration

Git is the most popular version control system in 2016. It enjoys a hefty 70% of all the search interest among these five VCSs.
Mercurial, with its 13.5% share, is still a tool of choice for companies like Facebook, Mozilla, Nginx, and NetBeans.
Subversion (SVN) has a 13.5% share as well, still used by some of the largest companies all over the world, such as Backblaze, FreeBSD, Mono, and SourceForge.
Perforce gained only 1.5% interest in 2016. It has been widely used by Google until they built their VCS called Piper in 2015.

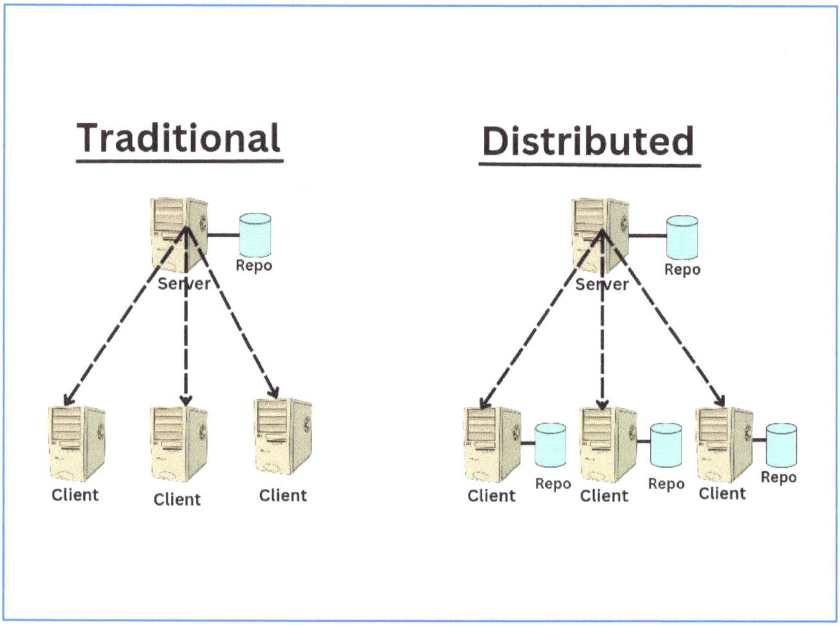

Figure: Difference between Traditional and Distributed VCS

The Version Control System Market is Segmented By Type (Distributed VCS, Centralized VCS), Deployment Mode (On-Premise, On-Cloud), End-user Industry (IT and Telecom, Retail & E-commerce, Healthcare and Life Sciences, BFSI), and Geography.

The above graph depicts that Git has been the most popular version control system. Git-related searches overtook the searches for Mercurial and CVS in 2007 and compete the biggest competitor Apache Subversion (SVN) in 2012. Since then, SVN has been evidence as second choice since in 2016. Perforce is striving to sustain but small share of interest.

Git is the leader in market and there is no doubt about it.If we refer to the overall search interest for these 5 version control systems, the search interest for Git has been growing constantly over time.

There were several reports published on internet on GIT Future scope. It is predicted that the global version control system market size was valued at USD 897.2 million in 2019 and is expected to grow at a compound annual growth rate (CAGR) of 12.6% from 2020 to 2026.

There are several organizations are migrating from traditional Version Control Systems to Advanced and Distributed Cloud-based Version control systems
It has been seen that many organizations have already migrated from SVN to GIT or have a good plan to migrate in the future. The total market share of Git is 77% as of now.

There are several reasons for organizations leaving SVN and adopting GIT.

Distributed in Nature
Git is distributed in nature. In case of catastrophic happens, like a server crash, every team member's local copy becomes a backup. It supports business continuity. SVN is centralized and entails a full dependency on a central server.

Staging Area and Local changes
Git facilitates developers to make several changes a day and some of them might not be a good candidate for the remote repository. Hence developers only move their changes to a special area called the staging area and keep other unwanted changes local without

affecting the staging area. You can push your changes to a remote repository only if arrives in staging and not directly. Git keeps this by using an indexing technique. SVN does not have this feature.

Merge Conflict Issues in SVN

SVN causes several merge conflicts whereas git facilitates you easily commit local changes before merging with a public branch

Branching Strategy

Branching in GIT is much easier and faster. Branching is complex in SVN.

Variety of Tools

Git has a rich set of a variety of tools to be used easily. These tools have been designed to keep day-to-day issues faced by developers in past.

Git has a variety of tools

Git Hooks

Git Blame

Git Cherrypick

Git Gitk

Git GUI

Offline Commits

Due to its design architecture git staging area keep all your project local. It helps developers to do offline build in case of travel or remote git is unavailable.

It means the developer can continue the development without having a dependency on the Git Remote server. SVN does not have this feature.

Chapter 13
DevOps and GIT

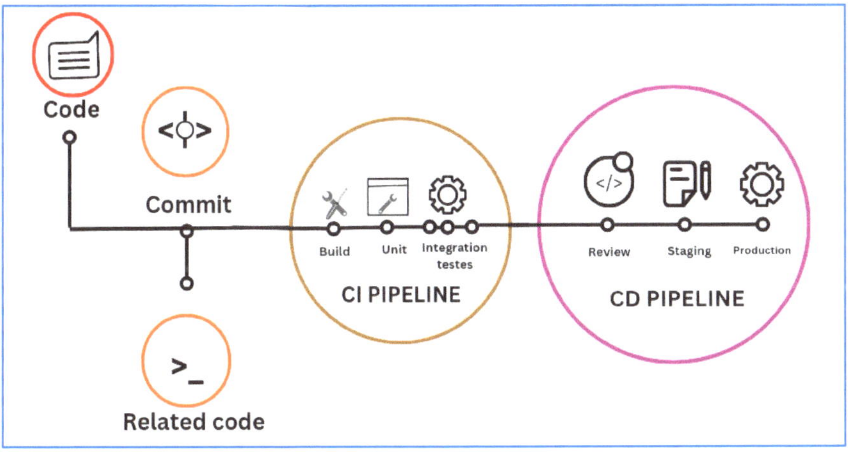

Figure: A typical view of CI/CD pipeline in DevOps.

DevOps is the combination of cultural philosophies, practices, and tools that increases an organization's ability to deliver applications and services at high velocity: evolving and improving products at a faster pace than organizations using traditional software development and infrastructure management processes.

When security is the focus of everyone on a DevOps team, this is sometimes referred to as DevSecOps.

13.1 Benefits of DevOps

Speed
Innovate for customers faster, adapt to changing markets better, and grow more efficiently at driving business results.

Reliability
You can reliably deliver at a more rapid pace while maintaining a positive experience for end users with reliable products.

Scale
Operate and manage your infrastructure and development processes at scale.

Improved Collaboration
Build more effective teams under a DevOps cultural model, which emphasizes values such as ownership and accountability.

Security
Move quickly while retaining control and preserving compliance.

DevOps Life Cycle

A DevOps process needs to use the right tools to address the key phases of the DevOps lifecycle.

These are three components of a DevOps lifecycle

Continuous Integration(CI)
Since developers make frequent changes, they can quickly spot problems (if any) and resolve them at an early stage.

Continuous Deployment(CD)

This process eliminates the need for scheduled releases and accelerates the feedback mechanism, allowing developers to address issues more quickly and with greater accuracy.

Continuous Testing

With the help of Automation tools, such as JUnit, Selenium, and TestNG, developed code is tested for bugs and errors that may have made their way into the code.

Continuous monitoring

Continuous monitoring is the process and technology used to detect compliance and risk issues associated with an organization's financial and operational environment.

13.2 DevOps Tools

DevOps tools help firms to resolve some of the challenges that come with the implementation of DevOps practices. However, there is no "one-size-fits-all" solution available out there. In the contemporary IT world, several variety of DevOps tools are available and fits for different requirement.

Version Control tools: GitHub, Bitbucket, GitLab

CI / Deployment Automation tools: Jenkins, Bamboo, IBM UrbanCode

Test Automation tools: Test.ai, Selenium

Container Management tools: Docker, Kubernetes, PCF

Performance Monitoring tools: Prometheus, Dynatrace, AppDynamics

Deployment & Server Monitoring tools;Splunk,Datadog

Configuration Management tools: Chef, Ansible, Puppet

Appendix

A. GIT CHEAT SHEET

GIT BASICS

git init <directory>	Create empty Git repo in specified directory. Run with no arguments to initialize the current directory as a git repository.
git clone <repo>	Clone repo located at <repo> onto local machine. Original repo can be located on the local filesystem or on a remote machine via HTTP or SSH.
git config user.name <name>	Define author name to be used for all commits in current repo. Devs commonly use --global flag to set config options for current user.
git add <directory>	Stage all changes in <directory> for the next commit. Replace <directory> with a <file> to change a specific file.
git commit -m "<message>"	Commit the staged snapshot, but instead of launching a text editor, use <message> as the commit message.
git status	List which files are staged, unstaged, and untracked.
git log	Display the entire commit history using the default format. For customization see additional options.
git diff	Show unstaged changes between your index and working directory.

UNDOING CHANGES

git revert <commit>	Create new commit that undoes all of the changes made in <commit>, then apply it to the current branch.
git reset <file>	Remove <file> from the staging area, but leave the working directory unchanged. This unstages a file without overwriting any changes.
git clean -n	Shows which files would be removed from working directory. Use the -f flag in place of the -n flag to execute the clean.

REWRITING GIT HISTORY

git commit --amend	Replace the last commit with the staged changes and last commit combined. Use with nothing staged to edit the last commit's message.
git rebase <base>	Rebase the current branch onto <base>. <base> can be a commit ID, branch name, a tag, or a relative reference to HEAD.
git reflog	Show a log of changes to the local repository's HEAD. Add --relative-date flag to show date info or --all to show all refs.

GIT BRANCHES

git branch	List all of the branches in your repo. Add a <branch> argument to create a new branch with the name <branch>.
git checkout -b <branch>	Create and check out a new branch named <branch>. Drop the -b flag to checkout an existing branch.
git merge <branch>	Merge <branch> into the current branch.

REMOTE REPOSITORIES

git remote add <name> <url>	Create a new connection to a remote repo. After adding a remote, you can use <name> as a shortcut for <url> in other commands.
git fetch <remote> <branch>	Fetches a specific <branch>, from the repo. Leave off <branch> to fetch all remote refs.
git pull <remote>	Fetch the specified remote's copy of current branch and immediately merge it into the local copy.
git push <remote> <branch>	Push the branch to <remote>, along with necessary commits and objects. Creates named branch in the remote repo if it doesn't exist.

GIT CONFIG

git config --global user.name <name>	Define the author name to be used for all commits by the current user.
git config --global user.email <email>	Define the author email to be used for all commits by the current user.
git config --global alias. <alias-name> <git-command>	Create shortcut for a Git command. E.g. alias.glog "log --graph --oneline" will set "git glog" equivalent to "git log --graph --oneline.
git config --system core.editor <editor>	Set text editor used by commands for all users on the machine. <editor> arg should be the command that launches the desired editor (e.g., vi).
git config --global --edit	Open the global configuration file in a text editor for manual editing.

GIT LOG

git log -<limit>	Limit number of commits by <limit>. E.g. "git log -5" will limit to 5 commits.
git log --oneline	Condense each commit to a single line.
git log -p	Display the full diff of each commit.
git log --stat	Include which files were altered and the relative number of lines that were added or deleted from each of them.
git log --author= "<pattern>"	Search for commits by a particular author.
git log --grep="<pattern>"	Search for commits with a commit message that matches <pattern>.
git log <since>..<until>	Show commits that occur between <since> and <until>. Args can be a commit ID, branch name, HEAD, or any other kind of revision reference.
git log -- <file>	Only display commits that have the specified file.
git log --graph --decorate	--graph flag draws a text based graph of commits on left side of commit msgs. --decorate adds names of branches or tags of commits shown.

GIT DIFF

git diff HEAD	Show difference between working directory and last commit.
git diff --cached	Show difference between staged changes and last commit

GIT RESET

git reset	Reset staging area to match most recent commit, but leave the working directory unchanged.
git reset --hard	Reset staging area and working directory to match most recent commit and **overwrites all changes** in the working directory.
git reset <commit>	Move the current branch tip backward to <commit>, reset the staging area to match, but leave the working directory alone.
git reset --hard <commit>	Same as previous, but resets both the staging area & working directory to match. **Deletes** uncommitted changes, and **all commits after** <commit>.

GIT REBASE

git rebase -i <base>	Interactively rebase current branch onto <base>. Launches editor to enter commands for how each commit will be transferred to the new base.

GIT PULL

git pull --rebase <remote>	Fetch the remote's copy of current branch and rebases it into the local copy. Uses git rebase instead of merge to integrate the branches.

GIT PUSH

git push <remote> --force	Forces the git push even if it results in a non-fast-forward merge. Do not use the --force flag unless you're absolutely sure you know what you're doing.
git push <remote> --all	Push all of your local branches to the specified remote.
git push <remote> --tags	Tags aren't automatically pushed when you push a branch or use the --all flag. The --tags flag sends all of your local tags to the remote repo.

Bibliography

Prana, G.A.A., Treude, C., Thung, F. et al. Categorizing the Content of GitHub README Files. Empir Software Eng 24, 1296–1327 (2019).https://doi.org/10.1007/s10664-018-9660-3

B. de Alwis and J. Sillito. 2009. Why are software projects moving from centralized to decentralized version control systems? In CHASE '09: Proceedings of the 2009 ICSE Workshop on Cooperative and Human Aspects on Software Engineering, pages 36--39, Washington, DC, USA. IEEE Computer Society

Alexander Yip, Benjie Chen and Robert Morris Pastwatch et al. a Distributed Version Control System,MIT Computer Science and AI Laboratory

Laadan, O., Nieh, J., Viennot, N.: Teaching operating systems using virtual appliances and distributed version control. In: Proceedings of the 41st ACM technical symposium on Computer science education. SIGCSE '10, New York, NY, ACM 480–484 (2010)

Clifton, C., Kaczmarczyk, L.C., Mrozek, M.: Subverting the fundamentals sequence: Using version control to enhance course management. SIGCSE Bull. 39(1) 86–90 (March 2007)

Sebastien Auvray, "Distributed Version Control Systems: A Not-So-Quick Guide Through" http://www.infoq.com/articles/dvcs-guide (last visited:2010/07/04)

Computer science curriculum 2008. Retrieved September 7, 2010 from
http://www.acm.org/education/curricula/ComputerScience2008.pdf

Index

A

B

C

D

E

F

G

S

T

U

V

W

www.ingramcontent.com/pod-product-compliance
Lightning Source LLC
LaVergne TN
LVHW051128210926
848482LV00001B/20

9798889253990